Digital Repertoires

Digital Repertoires

Embedded and everyday technologies in later life

Edited by Riitta Hänninen, Sakari Taipale and Laura Haapio-Kirk

UCL PRESS

First published in 2025 by
UCL Press
University College London
Gower Street
London WC1E 6BT

Available to download free: www.uclpress.co.uk

A CIP catalogue record for this book is available from The British Library.

ISBN: 978-1-80008-843-6 (Hbk)
ISBN: 978-1-80008-845-0 (Pbk)
ISBN: 978-1-80008-844-3 (PDF)
ISBN: 978-1-80008-846-7 (epub)
DOI: https://doi.org/10.14324/111.9781800088443

Printed and bound by CPI Group (UK) Ltd, Croydon, CR0 4YY

Contents

List of figures

List of contributors

Earvin Charles B. Cabalquinto is an ARC DECRA Research Fellow and Senior Lecturer in the School of Media, Film and Journalism at Monash University, Melbourne. He is the author of *(Im)mobile Homes: Family life at a distance in the age of mobile media* (Oxford University Press, 2022). His expertise lies in the intersecting fields of digital media, mobilities, migration and ageing research. His current projects critically examine the experiences, dynamics and consequences of digital exclusion among migrants and their support networks through a global perspective.

Marília Duque has a PhD in Communication Studies (ESPM São Paulo), is a researcher at the LADIGE Digital Lab for Food Education and Humanities (Federal University of Rio de Janeiro) and a research associate on the Anthropology of Smartphones and Smart Ageing Project (University College London).

Laura Haapio-Kirk is a Junior Research Fellow in Social Anthropology at Christ Church, the University of Oxford. Her research examines the intersection of experiences of ageing, digital practices and wellbeing, based on fieldwork conducted in Japan and Finland. She is the author of *Ageing with Smartphones in Japan: Care in a visual digital age* (UCL Press, 2024). Laura is currently exploring graphic modes of research and dissemination. She was a Leach Fellow in Public Anthropology at the Royal Anthropological Institute of Great Britain and Ireland

Mikaela Hellstrand is a PhD student in Health Informatics and Logistics, at the Department of Biomedical Engineering and Health Systems, KTH Royal Institute of Technology, Sweden. Her current research explores how care is reconfigured with the introduction of data-driven technologies in the context of care for the elderly in Sweden. With a background in anthropology, she has a particular interest in experimenting with, and reflecting upon, ethnography and ethnographic methods.

Riitta Hänninen is an Adjunct Professor and Senior Researcher at the Centre of Excellence in Research on Ageing and Care in the Department of Social Sciences and Philosophy, University of Jyväskylä. Her research focuses on ageing, digitalisation, digital inclusion and digital ethnography. Dr Hänninen's previous work has been published in several peer-reviewed anthologies and journals such as *New Media & Society*; *Information, Communication & Society*; *Leisure Studies* and *Media & Communication*.

Loredana Ivan is a Professor at the National University of Political Studies and Public Administration, Communication Department, Bucharest, Romania. She is PI for the Older Adults Co-Creating a Sustainable Age-friendly City project and a co-applicant in the project Aging in Data, financed by the Social Sciences and Humanities Research Council of Canada. Her areas of interest include interpersonal communication mediated by technology, fake news and generations, family communication and ageism. She is an editor of the recent special issues *AI in the Everyday Life of Older Adults: Panacea or Pandora's Box* and *Post-pandemic Digital Realities of Older Adults*.

Sanna Kuoppamäki is an Assistant Professor in Technology in Health Care, Department of Biomedical Engineering and Health Systems, KTH Royal Institute of Technology, Sweden. Her research explores the use and design of interactive technologies for health and wellbeing from the life course perspective. She is a co-PI of the project 'Advanced Adaptive Intelligent Systems' (Digital Futures).

Donald McMillan is a Senior Lecturer at the Department of Computer and System Sciences, Stockholm University, Sweden. His research explores the impact of novel technology and multimodal interaction mechanisms on everyday work and home life. He is a co-PI of the project 'Advanced Adaptive Intelligent Systems' (Digital Futures).

Daniel Miller is Professor of Anthropology at UCL. He is director of the ASSA project and was director of the Why We Post project. He is the author/editor/co-author of 43 books including *The Good Enough Life*, *Ageing with Smartphones in Ireland* (with P. Garvey), *The Global Smartphone* (with ten others), *How the World Changed Social Media* (with eight others), *Social Media in an English Village*, *Tales from Facebook*, *Digital Anthropology* (edited with H. Horst), *The Comfort of Things*, *Stuff*, *A Theory of Shopping* and *Material Culture and Mass Consumption*.

Sakari Taipale is Professor of Social and Public Policy at the University of Jyväskylä, Finland, and the head of New Technologies, Ageing and Care research group at the Centre of Excellence in Research on Ageing and Care (CoE AgeCare). He is author/co-editor of the books *Intergenerational Connections in Digital Families, Digital Technologies and Generational Identity: ICT usage across the life course* and *Social Robots from a Human Perspective*.

Emilene Zitkus is an Associate Professor in Inclusive Design at Loughborough University. Her research interests include human-computer interaction, digital inclusion, user-centred and user-experienced design. Her expertise in inclusive design has been honed during the last 15 years through projects developed with industrial partners such as Ford and Jaguar-Land Rover and research funded by the Royal Society, EPSRC, FAPESP, CNPq, CAPES and the Information Design Association.

Acknowledgements

This book is about the invisible, embedded technologies that wield the greatest impact on people, social relationships and societal functioning over time. Without the co-operation and commitment of our research participants, it would be very difficult to comprehend the full impact of these mundane technologies, both devices and applications, that are the most transformative of everyday life. Producing this book has been a rewarding journey into the everyday lives of older people across the globe, and we thank them very much for their support and collaboration.

We have brought together researchers from different countries, representing four continents, to explore the digital repertoires of older people around the world, including their daily digital practices that make technology meaningful to them. We are deeply grateful to all the contributing authors for their insight and commitment to our joint project. All the chapters included in this volume represent the professionalism of the authors and respect for scientific inquiry and research ethics. We would like to thank all the participants of the workshop 'Embedded and everyday technology – Digital repertoires in an ageing society', which took place on 28 January 2022. Many of you contributed a chapter to this book, and all of you participated in discussions and provided feedback that shaped the direction of our work. The seeds of the book were sown in this hybrid format workshop held at the University of Jyväskylä, Finland. The workshop was part of the activities of the Research Council of Finland's Centre of Excellence in Research Ageing and Care (CoE AgeCare) and Towards Socially Inclusive Digital Society (DigiIN) project.

We would also like to thank all the reviewers who meticulously reviewed our chapters and have allowed their names to be published here: Michela Cozza, Marija Djurdjevic, Vera Gallistl, David Prendergast, Susanna Rivinen, Jane Vincent, Shireen Walton and Barbara Zach. Thank you sincerely for your time, expertise and constructive feedback. Your rigorous assessments have enhanced the quality of this volume. Thanks also to Charlotte Linton for her comments on Chapter 5.

Our thanks go to Professor Daniel Miller for providing such an insightful epilogue to the book. He also brought some of the authors together at the Anthropology of Smartphones and Smart Ageing (ASSA) conference in London, UK, 8 –11 June 2022. We would also like to thank Professor Sirpa Wrede for her encouragement to delve into the concept of digital repertoire and Project Researcher Sini Tiihonen for her invaluable help with finalising the book.

In uncovering the significance of seemingly mundane technologies in the everyday lives of our research participants, this volume aims to highlight their profound influence on interpersonal relations and experiences of ageing. We hope that we have done justice to the stories shared with us, and that this collective effort contributes to a deeper understanding of the complexities surrounding technology adoption and usage among older populations. Through the diverse perspectives shared, we aspire to inform future research and societal efforts aimed at enhancing digital inclusion and accessibility for all generations.

Part I

Re-conceptualising digital heterogeneity

1

Introduction: older adults navigating digital technologies

Sakari Taipale, Riitta Hänninen and Laura Haapio-Kirk

This book explores the embeddedness of digital technologies in later life, touching on two societal megatrends reflected in the everyday lives of older adults: digitalisation and ageing. Both global trends are incredibly recent, as for most of human history people have lived relatively short lives. The current cohort of older adults are the first in history to experience the rise of digital technologies in their daily lives, thus presenting an invaluable opportunity to question the presumed convenience, affordability and availability of such technologies – for example, personal mobile devices, social media platforms, applications and electronic services – which now feature in several governmental strategies that promote digitalisation for everybody (Cozza et al. 2017; Gallistl et al. 2020; Selwyn 2004). The simultaneous emergence of societal ageing as an irreversible global trend challenges the supposed ubiquity and homogeneity of digital technologies and their users.

Since the late 1990s, when digitalisation started to gain a stronger foothold in developed societies, one of the key questions has been how to engage older adults with various technologies and to ensure their digital inclusion (e.g. Hargittai 2002; Hänninen et al. 2021a; Kuoppamäki et al. 2022; Seifert and Rössel 2019). The apparent difficulty of this question lies not only in the heterogeneity of older people, but also in the constant technological change and design principles that fail to recognise sufficiently the needs associated with later life (Pajula et al. 2024). While people are living longer, the lifespan of new digital technologies is still very short. This means that people have to adopt and adapt to evolving technologies at an increasingly older age.

In their everyday lives, people face this reality from very different starting points. Older adults come from varied backgrounds with differing resources, needs, skills and motivations to adopt digital technologies into their daily lives (Fernández-Ardèvol et al. 2017; Friemel 2016; Hänninen et al. 2020). While many older adults view digital technologies and services in a positive light, others do not want to engage with them – or, due to various age-related reasons, for example poor physical condition or cognitive issues, are not able to benefit from digital technologies to the same degree as other age cohorts (Peek et al. 2016; Quan-Haase et al. 2017). Furthermore, it is worth noting that older adults cannot be reduced solely to their chronological age. Ageing as such is no obstacle to active and independent use of digital technologies.

The relationship between older adults and their digital technologies is developed in the interplay of various cultural and socio-demographic factors such as gender, language, ethnic background and income (Nimrod 2017; Olphert and Damodaran 2013; Scheerder et al. 2017; Tsatsou et al. 2018). As views of ageing and age stereotypes vary across countries and cultures (P. de Paula Couto, et al. 2023), it is also a vastly diverse experience to be an older technology user in different parts of the world. The role that technology plays in the daily lives of older people around the world, the meanings that they attach to it, the normative expectations that communities have of older people's use of technology and the policies that have been formulated to regulate its use all vary widely.

In previous studies on digital technology use by older adults, starting points of the research have ranged from questions focused on access (Czaja and Lee 2007) and skills (Bhattacharjee et al. 2020; Hargittai et al. 2019) to more comprehensive accounts of the benefits and meaning (Bossio and McCosker 2021; Hill et al. 2015; Hänninen et al. 2023a) of digital technologies in later life. A strong focus on the technological perspective of digital technology use, particularly in early research on access and ageing, underlined the significance of devices and infrastructure as key components of digital inclusion. Similar technological emphasis can also be found in studies on digital skills and competence, in accordance with which digital skills are frequently portrayed as something that older adults lack (van Deursen et al. 2011; Loos and Ivan 2022), rather than focusing on the more complex question of what it signifies for older adults to use – or not to use – digital technologies, as well as the roles digital technologies play in their daily lives and the meanings attached to them.

In this book we explore the embeddedness of digital technologies in the everyday lives of older adults. By zooming in on the everyday practices of digital technology, we aim to unfold some of the mundane, recurrent and unexpected factors that facilitate older adults' participation in the digital society. The concept of 'everyday life' is known to be an elusive one; scholars have attempted to define it since the term become popular in the 1920s (Bennett and Watson 2002; Ebrey 2016). In short, it refers to the mundane, ordinary and seemingly unremarkable aspects of human existence. Everyday life brings together the daily routines, habits and interactions that make up our lives, and that provide security and continuity in turbulent times (Nyman 2021; Pink 2012). Within everyday life are the hidden meanings and significances embedded in digital practices, constituting the digital age, and indeed the very fabric of society.

The concept of the mundane offers advantages for studying the role and behaviour of older adults in the digital society. First and foremost, it places people, as opposed to technology, at the centre of analysis. Second, it has also been shown that the most mundane activities contain potential for resistance, creativity and social change (de Certeau 1984; Lefebvre 1947). It is easy to imagine how older adults' reluctance to use various digital technologies can be seen as a silent protest against the digitally saturated modern world designed by young people (often men) for young people (see Lewis et al. 2006). However, this book also highlights instances in which older adults' digital practices involve a lot of creativity, helping them to meet their daily needs. Everyday life is thus a kind of 'comfort zone' that always allows for innovation and unusual practices (Hartman 2008). In this sense, everyday life has proved itself to be concerned with not only ordinary things, but also the extraordinary choices, strategies and negotiations that people engage in to preserve their wellbeing and the comforts of their daily existence.

The aim of the book

Reflecting on the overall development of digitalisation as a societal process, there is a clear gap in previous theoretical discussion on the individual perspective of digital technology use in an ageing society. To work towards this end, in this book we explore the notion of digital repertoire, which consists of the routine practices individuals typically develop when using digital technologies and services (Stevenson 2013, 156). Later in this chapter we outline the contours of the digital repertoire based on previous literature. Given that there are no widely accepted and applied definitions of digital repertoire at hand, this book aims to clarify the concept's meaning and to demonstrate how it can be applied to understand the heterogeneity characteristic of technology use in later life.

The concept of digital repertoire reveals how older adults' digital participation is less about ensuring access and learning technical skills and more about understanding how older people adapt their needs around digital technologies, at the same time shaping these technologies. A digital repertoire approach asks both what is required for older adults to use various digital technologies and, consequently, why they choose or otherwise end up incorporating certain devices, applications and services into their daily lives. The chapters explore the different ways in which older people navigate an increasingly digital everyday life, drawing on their personal digital repertoires. Through a series of case studies, the chapters show how older adults react to rapidly evolving technologies and how in many cases they overcome barriers by using a socio-technical toolkit we call their digital repertoire.

The concept of repertoire has been fruitful in social theory for several decades, starting with Ann Swidler's work on 'cultural toolkits' (Swidler 1986). In Cati Coe's work on the cultural repertoires of Ghanaian transnational families, she helpfully defines the contours of repertoires, which we argue can be applied equally to digital repertoires: they imply multiplicity; they are both habitual and conscious; they are enacted; they have dimensions of power and status; they can be reshuffled and transposed (Coe 2013). Coe advocates for repertoire over the linked notion of habitus (Bourdieu 1977 [1972]; Mauss 2006) by highlighting how it

> signals a multiplicity of cultural resources and frameworks, a body or collection of practices, knowledge, and beliefs that allows people to imagine what is possible, expect certain things, and value certain goals. (Coe 2013, 15–16).

Habitus, in contrast, places emphasis on often unconscious dispositions established during early experiences within the family. In addition, repertoire is open to change, rather than being fixed over a life course as habitus tends to be (Coe 2013, 16).

Digital Repertoires identifies a variety of mundane and embedded digital practices that underpin older adults' participation in digital societies. It is these routine-like practices, often taken for granted, that provide continuity and stability in ever-changing technological environments. Digital participation is not only enabled by the provision of new technologies and basic infrastructure, nor by formal digital skills training. Often more important for older adults' digital participation are self-initiated, mundane and unobtrusive practices that help to overcome unexpected technical problems (Hänninen et al. 2022). They may involve support provided by 'warm' experts, such as family and friends, and the sharing of resources, or manifest themselves as 'workaround' practices that help people to overcome the problem they have encountered (e.g. Hänninen et al. 2021b; Korpela et al. 2024).

The chapters in this book include three approaches to digital repertoires in the context of everyday technology use in later life. First, the book focuses on unravelling the fundamental elements of digital repertoires through both theoretical and empirical analysis. Second, the book provides several ethnographic examples from around the world to illustrate the embedded digital practices reflecting digital repertoires. These examples address the cultural, social and individual aspects of using digital technologies in later life. Third, in addition to the ethnographic narratives of digital practices, the book offers views on how mundane digital practices in the everyday lives of older adults, as well as visual methods in documenting them, can inform the design of digital technologies.

By adopting this approach, the book develops a critical and multidisciplinary perspective on the heterogenous use of digital technologies and the relationship between actual uses and design practices. In doing so it challenges various stereotypical views of old age as a homogeneous category or an age cohort-based construct. All three perspectives are crucial in attempting to bring older adults forward as protagonists of their digital lives, rather than mere recipients of digital technologies. Ultimately *Digital Repertoires* also contributes to a fuller definition of the digital repertoire in connection to an ageing society.

Defining the concept of digital repertoire

In this book we introduce the concept of digital repertoire to the study of technology and ageing, with the intention of unfolding the nuances and depth contained in the heterogeneity of digital later life. We first developed the concept while analysing interview data on the digital daily lives of Finnish older adults (Hänninen et al. 2021b). At that time the concept emerged somewhat endogenously from the research material to describe the great diversity of everyday practices that older adults had developed, in order to manage all the digital devices and services that were important to them if not essential. Based on this analysis, which initially tackled the question of what it takes for older adults to adopt and use various digital technologies, we found the answer to be deeply embedded in the individual and social, or shared, aspects of each older adult's personal digital repertoire.

Shortly afterwards we became aware that the concept had been used in earlier studies. This usage was often without a clear definition, however, or concentrated on a single media use (Nimrod 2017). It was used to make sense of the strategies and methods people adopt while employing a certain technology in various contexts (for example, on linguistic practices, McLaughlin 2014; contention repertoires, Mare 2016; digital network repertoires Chadwick 2007). A more serious attempt to define digital repertoire and to use it as an analytical concept was developed by Stevenson (2013) who, in his study on classroom learning, noted that 'digital repertoires enable individuals to both use technologies fluently and adapt their knowledge to new situations and contexts' (150). While prior studies presented either a highly instrumental or technologically oriented approach to digital repertoires, it was clear that a need existed to explore the concept more holistically. We considered that the concept has significant potential in unfolding the diversity of older people's technology use.

To provide a common starting point for the book and its contributing authors, a working definition of the digital repertoire was further developed as the chapters were drafted. We suggested that the concept of digital repertoire encapsulates the variety of ways in which people make their personal technologies meaningful to them through everyday practices. The authors were guided to use case studies to explore how older adults adopt and use digital technologies along the lines of their personal digital repertoires. The digital repertoire was presented to the authors as consisting not only of a combination of interests, skills and availability of devices and applications, but also of a user's wider networks of support and services (Donner 2015; Hasebrink and Domeyer 2012, Hasebrink and Popp 2006; Hänninen et al. 2023b). We have invited the authors to build on this initial concept by taking a broad yet intimate look at the ways in which older adults manage their digital devices, applications and digital information in everyday life.

Situating the concept of digital repertoire

To date, disparities in older adults' digital technology adoption and use have been studied through multiple frameworks, such as technology adoption models, digital divides approach, as well as uses and gratification and ageing theories. Digital repertoire may easily be engulfed in this wide sea of theoretical frameworks and closely related concepts that address the inequalities and opportunities related to participation in digital societies. It is therefore worth considering its specificity and value in relation to its adjoining conceptual and research traditions. While our aim here is briefly to situate the concept within a wider research tradition of digital inequalities and participation, more systematic accounts of the theoretical underpinnings of the digital repertoire concept (Hänninen and Taipale, Chapter 3) and analysis of its differences to neighbouring concepts (Ivan, Chapter 2) are discussed separately.

The digital repertoire refers to the individual combination of resources and properties available to a person navigating the ocean of digital devices and services. The digital repertoire is therefore by definition unique to everyone, regardless of age, gender or other cultural or socio-demographic factors. On the one hand, this uniqueness calls for an ethnographic approach that facilitates delving into everyday life, mundane practices and the embedded meanings ascribed to digital technologies. Furthermore, digital repertoire also consists of social, or shared, characteristics that underline the significant contribution of family, friends and social networks to this individually structured collage of resources and properties (Hänninen et al. 2023b).

On the other hand, the individuality of digital repertoire also distinguishes it from other approaches (digital divides, digital gaps, digital skills) that aim to group people into separate, measurable and governable categories according to their access to technology, digital skills or ways of using new devices and services (Barzilai-Nahon 2006; Vehovar et al. 2006). For example, the concept of 'digital divide' was built on the idea that there are differences in usage and skills between certain groups of people; the groups that are most 'vulnerable', such as women, older people or those with lower levels of education and/or income can thus be detected (Helsper and van Deursen 2017).

However, as such groups have recorded the greatest increase of internet users in recent years – although some digital divides have also deepened (Blank et al. 2020; Pérez-Amaral et al. 2021) – a rethink is required to study the ways in which people in all their heterogeneity are using digital technologies today. Measuring digital divides also made it possible to examine whether such gaps led to differences in levels of wellbeing and satisfaction with life (Büchi et al. 2018; Livingstone et al. 2023; Rosenberg and Taipale 2022). While these kinds of conceptual approaches may successfully feed back into policy-making, and positively influence the allocation of public resources for those in need, they tend to portray older people as a thoroughly 'disadvantaged' group, lacking the highest levels of digital literacy or access to the latest communication technologies. The heterogeneity of old age is widely overlooked.

From a critical standpoint it can be argued that, in comparison to the concept of media repertoire, digital repertoire is primarily concerned with the digital aspects of everyday life, thereby neglecting the non-digital. While this perspective is undoubtedly pertinent when considering the digital repertoires of older adults, it does not necessarily follow that the concept of a digital repertoire is insensitive to the non-digital aspects of everyday life. On the contrary, as has been demonstrated in previous studies (Hänninen et al. 2023a; Hänninen et al. 2023b), the adoption and use of various digital technologies typically occur within the context of what is personally meaningful or relevant to someone, irrespective of whether the subject of interest or practice is inherently digital. This is particularly the case where older adults are integrating novel digital technologies, such as new applications and platforms, into their daily lives. Moreover, digital skills and competencies are often augmented and reinforced by non-digital alternatives. For instance, individuals may opt to book a doctor's appointment by telephone instead of using digital services, or they may 'piggy-back' on their spouse's usage to connect with family and friends on social media (Haapio-Kirk 2024, 58–9).

Such scenarios also fall within the purview of the concept of digital repertoire. As this book will demonstrate, digital repertoires encompass not only digital competencies and practices characteristic of ageing, but also many non-digital aspects of everyday life. This in turn reflects the varying contexts, skills and forms of co-use, proxy use and non-use of digital technologies in later life. Rather than suggesting that the complexities of these interactions be oversimplified by conflating diverse experiences and contexts under a single, broad concept, we see the broadness of 'digital repertoire' as a strength, inviting inclusive and flexible analysis that can adapt to the varying ways in which individuals integrate technology into their lives. Particularly in the case of older adults, who may not consider themselves 'expert' at digital technologies, the concept affirms the diversity of digital repertoires for everyone.

Another value of digital repertoire as a concept is the qualitative emphasis, which encompasses the everyday life needs, interests and situations encountered by older adults. The concept of digital repertoire is, in many ways, a contextualising concept. As noted above, it refers to the ability to describe the complex whole of everyday life practices and the overlapping and complex relations of various digital devices, applications, services and networks, both in terms of the 'digital' and the 'non-digital'. Although these elements are often challenging to quantify or measure, the concept of digital repertoire can assist in identifying phenomena and patterns that can subsequently be studied in quantitative terms.

Although the concept of digital repertoire acts as a common theme in this book, each chapter explores and develops the concept based on individual theoretical and empirical foci. In Chapter 3 Hänninen and Taipale attempt to situate the digital repertoire in relation to three major theoretical approaches: 'digital divide theory', 'practice theory' and the 'uses and gratification theory' (UGT). They also demonstrate that the concept of digital repertoires brings together the unique combinations related to the personal (individual), shared (social) and cultural aspects of digital technology use in the contexts of everyday life. By investigating the cultural, social and individual aspects of digital technology use, the authors of other chapters challenge stereotypical understandings of 'old age' as a homogenous category, or the existence of a construct based on age-cohort. In this light the notions of digital repertoire and heterogeneity serve as connective agents in the overall structure of this book.

The structure of the book

Each chapter in this volume contributes to the overall aim to examine the everyday digital practices of older people around the world and understand how their cultural, shared and individual contexts shape the meaning they bestow upon technologies. The chapters reveal the complex relationship between digital technologies and experiences of ageing; they also highlight the cultural differences and similarities in the digital repertoires of older people across the world. The book is divided into three parts, all with a slightly different focus. The chapters in Part I guide the reader from a broad theoretical starting point towards Part II, in which empirical case studies show how digital repertoires are manifested in practice in diverse settings. Part III builds upon the themes and concepts discussed in the preceding parts, showing how they can be combined in, and influence, processes of design. The epilogue provides commentary on the usefulness of digital repertoire as a lens not only for scholars of digital culture, be they anthropologists, sociologists, science and technology scholars or any number of associated fields, but also for designers, policy-makers and practitioners working in the urgent and growing area of digital inclusion.

Part I, 'Re-conceptualising digital heterogeneity', consists of two chapters outlining the theoretical roots and conceptual contours of digital repertoire and the significance of heterogeneity and the notion of mundane in older adults' technology use. It focuses on the central conceptual aspects of digital repertoire and examines it in the context of theoretical discussions of older adults' digital technology use, digital divides and digital skills.

In Chapter 2, 'From digital skills to digital repertoire: towards a more inclusive conceptualisation of technology use in an ageing society', Loredana Ivan argues that in comparison with digital skills and skills-related divides, the concept of digital repertoire provides a more inclusive take on technology use. In particular it highlights the individual and shared (social) aspects of digital technology use. The more common concept of 'digital skills' is not only hegemonic, considering how different social groups use technology, but it also presents a Cartesian duality of 'advanced' technology users as opposed to the 'laggards'. Chapter 2 discusses digital repertoire as an inclusive and nuanced concept, emphasising the ways in which different social groups use communication technologies in everyday life. Ivan argues that digital repertoire is not an immobile concept but rather one that encompasses the inherently processual nature of digital technology use. She explores the implications this has on digital policy-making for older adults, who come from all walks of life and are themselves ever-changing as they age.

These ideas are further developed by Riitta Hänninen and Sakari Taipale in Chapter 3, 'Exploring the key elements of digital repertoire: the meaning of new technology in later life'. In this chapter the authors map out the elements that constitute the digital repertoire of a given older adult, considering what, in fact, it takes for older adults to build a meaningful relationship with digital technologies. The notion of digital repertoire is defined in this chapter through the meaningful relationships developed by older adults in their everyday lives within a specific socio-cultural context. Drawing from a Finnish case study, the chapter positions the concept of digital repertoire in relation to social theories commonly applied in the field of ageing and digital technology studies. It identifies three individual and three social, or shared, elements of digital repertoire in later life, all manifested through the everyday life practices of older adults. The analysis also describes how the individual and shared aspects of digital repertoire reflect wellbeing in later life. It provides a stepping stone towards Part II, where digital repertoire is further explored in two ethnographic case studies.

Part II, 'Digital repertoires in everyday life', investigates the mundane and embedded digital practices in the context of care and digital technologies. Chapters 4 and 5 both contribute to the definition of digital repertoire by providing culturally specific results that reflect the characteristics of digital repertoire across various geographical contexts. In Chapter 4, 'Lives on hold? Ageing migrants' affective digital repertoires during lockdowns in Victoria, Australia', Earvin Cabalquinto explores older migrants' digital practices in order to understand their everyday digital repertoires during the Covid-19 pandemic in Australia. The chapter reveals the ways in which lockdown-caused immobilities are managed and surpassed via various digital technologies. The ethnographic analysis also underlines the affective elements of digital repertoire, including happiness, belonging, frustration and ambivalence, emerging from accessing and using digital technologies for forging local and transnational connections.

Furthermore, Chapter 4 provides nuanced perspectives on the heterogenous digital practices of ageing migrants in Australia by examining how diverse motivations and outcomes of affective connection are influenced by disproportionate socio-cultural infrastructural systems. In doing so, Cabalquinto sheds light on the politics of digital repertoires, revealing how both connective and disconnective practices underline the continued negotiation of structural and digital inequalities of contemporary life.

Part II closes with Chapter 5, 'Digital repertoires of care in Japan: a participatory visual approach', by Laura Haapio-Kirk. This chapter brings the visual aspects of digital repertoires to the discussion. Based on her extensive fieldwork in Japan (Haapio-Kirk 2024), the author explores the pivotal role of embedded visual communication practices in older people's everyday lives. Through smartphone-based elicitation interviews and comics-based graphic ethnography, this chapter presents evidence that visual digital communication is becoming an increasingly important part of the digital repertoires of older, particularly female, Japanese smartphone users. Two case studies reveal how emerging digital repertoires allow older adults to mitigate their sense of vulnerability and dependency on others as ageing persons, and to deflect negative associations around ageing that have emerged in discourses about Japan's 'super ageing' society. Digital repertoires are shown to be highly personal, based on individual life trajectories, yet at the same time they are implicitly social, cultivated within and between cultural norms.

Part III, 'Mundane practices and visual methods', encompasses Chapters 6 and 7. These two chapters explore how routine-like practices, and the use of visual methods to study them, can inform the design of digital technologies and services, and further unpack the diversity of user behaviours contained in their digital repertoires. While previous chapters presented the key elements of the digital repertoire as a concept and as everyday practice, this section highlights the benefits and challenges of the concept for informing the design processes and co-creation of digital services – where the focus is usually much narrower, covering only accessibility, usability and technical skill requirements. The main idea of the third part of the book is to demonstrate that the concept of digital repertoires and visual methods helps to uncover the complexities of later life heterogeneity, a prerequisite for designing more inclusive and personalised digital services.

Chapter 6, 'Ageing in digitalised Brazil: using the 'Brazilian way' to compensate for lower digital skills', by Marília Duque and Emilene Zitkus, provides an ethnographic study of mundane digital health practices in Brazil. Duque and Zitkus address the ways older people in São Paulo, Brazil, engage with digital technologies and their online experiences, the devices older adults use and the support they have on daily basis. The study shows how older adults resort to their social networks via WhatsApp to gain access to health services that are not always readily available for them.

These informal connections, which constitute an integral part of older Brazilians' digital repertoires, are also used to compensate for lack of digital skills and to search for health information online. Together, they provide a powerful informal network based on favours, or 'the Brazilian way', that help older adults to deal with bureaucracies and constraints in their everyday lives. Chapter 6 also adds to our understanding of how the design of public digital services could benefit from subscribing to these informal digital networks and practices to support the inclusion of older people as digital citizens.

In Chapter 7, 'Uncovering older adults' digital repertoires for designing conversational scenarios: graphic transcript as a design method', Sanna Kuoppamäki, Mikaela Hellstrand and Donald McMillan explore how the design process of a conversational agent could be improved to better acknowledge older adults' mundane digital practices and diverse digital repertoires. The authors investigate how video-ethnomethodology and graphic transcripts can inform the design process of conversational agents, such as Google Home or Amazon Alexa, by uncovering the mundane practices of everyday life in a specific domestic task, such as cooking a meal at home.

In Chapter 7 the analysis builds on the discussion of graphic methodology introduced in the previous section to show how visual methods can reveal digital repertoires as part of design processes. The application of graphic transcripts highlights older adults' physical and spatial interactions, social and collaborative tasks, temporalities and interconnected actions for the purpose of designing voice-based interaction systems in the kitchen. The notion of digital repertoire also provides a conceptual lens for design and co-design of interactive technologies with older adults, and contributes to more inclusive and sensitive human-robot interaction.

Conclusion

In this book we have developed the concept of digital repertoire to cover a stock of difficult-to-identify skills, daily practices and attitudes that go beyond the use and non-use of technology or specific digital skills. The strength of digital repertoire, which seeks to shed new light on the heterogeneity of digital technology use in later life and beyond, is based on its holistic scope and applicability. Previous concepts, by contrast, can only partially explain a variety of digital technology uses in everyday life contexts.

This book proposes that older adults exhibit a digital repertoire characteristic to their personal daily life. Moreover, as discussed in Chapter 3, there are certain key elements to these digital repertoires that are typical of older adults. Instead of focusing on specific skills, devices or applications, older adults view digital technology especially in terms of its meaning and function in their daily lives. Communication technologies can be highly affective, connecting families across distances and decreasing isolation, as Chapter 4 illustrates. Challenges to mobility are not limited to the experiences of older adults, of course; we have all witnessed restrictions imposed by a global pandemic, prompting novel reiterations of our digital repertoires.

Throughout this volume, the concept of digital repertoire emerges as both a dynamic phenomenon, fluid and responsive to the needs and wishes of older adults in their digital lives, and a culturally nuanced set of practices; Chapter 5, for example, reveals Japanese communicative norms. As highlighted in Chapter 6, what constitutes a meaningful digital practice is deeply influenced by cultural contexts, such as the so-called 'Brazilian way', where behaviours considered mundane in one setting may be extraordinary in another. The chapters that follow underscore the heterogeneity of older technology users and correspondingly their digital repertoires, emphasising the need for a nuanced understanding of digital practices across diverse cultural landscapes.

Limitations in digital resources and skills can provoke new and innovative ways of tuning in with digital technologies and various online environments, and of making them one's own by older adults. The heterogeneity discussed in the book thus entails not only the idea of what is characteristic to older adults as a group of people belonging to a specific age cohort, but also, and especially, the personal, social and cultural idiosyncrasies of digital repertoire in later life. In terms of applicability, it is also worth noting that the notion of digital repertoire is not something that comes with ageing. It is rather something that we all have as digital technology users.

The methodological innovation demonstrated in this book offers valuable pathways for further exploration of digital repertoires. Through approaches such as investigating visual communication and employing graphic transcripts of video records, researchers can delve deeper into the complexities of digital practices and their cultural underpinnings.

The last part of this book asks an important question: how should digital repertoires be incorporated into the design of digital technologies? Technological design often assumes that everyone has the same abilities or cultural values. If designers are able to recognise the varied digital repertoires of older people, and the values they represent, they can create better future technologies for everyone. It is imperative for designers to go beyond considerations of accessibility and usability, and to instead strive to incorporate the values and shared practices that shape individual ways of using technology. By doing so, designers can create more inclusive and culturally responsive digital environments that empower users to engage meaningfully with technology, in ways that resonate with their lived experiences and cultural backgrounds.

In conclusion, as we reflect on the exploration of digital repertoires presented in this volume, it becomes evident that our journey is ongoing. There exists a pressing need for continued research, collaboration and advocacy efforts to ensure equitable access and meaningful participation in the digital world for older people. As you read this volume we invite you to reflect on how your attitudes and behaviours shape your own digital repertoires, and how both individual and shared repertoires change over time. It is our hope that such reflections may provoke consideration on how we might contribute to the creation of age-friendly digital spaces within our communities and organisations. Ultimately this collective endeavour, spanning research, design and policy, is based on a commitment to digital inclusion that holds the power to affect positive change across generations as digital technologies become ubiquitous throughout the world.

References

Bakardjieva, M. 2005. *Internet Society: The internet in everyday life*. London: Sage Publications Ltd. https://doi.org/10.4135/9781446215616.

Barzilai-Nahon, K. 2006. 'Gaps and bits: Conceptualising measurements for digital divide/s', *The Information Society* 22(5): 269–78. https://doi.org/10.1080/01972240600903953.

Bennett, T. and D. Watson (eds). 2002. *Understanding Everyday Life*. Oxford: Blackwell.

Bhattacharjee, P., S. Baker and J. Waycott. 2020. 'Older adults and their acquisition of digital skills: A review of current research evidence', *Proceedings of the 32nd Australian Conference on Human-Computer Interaction*, 437–43. https://doi.org/10.1145/3441000.3441053.

Blank, G., W. Dutton and J. Lefkowitz. 2020. *OxIS 2019: Digital divides in Britain are narrowing but deepening*. University of Oxford, Oxford: Oxford Internet Institute. https://ssrn.com/abstract=3522083.

Bossio, D. and A. McCosker. 2021. 'Reluctant selfies: Older people, social media sharing and digital inclusion', *Continuum* 35(4): 634–47. https://doi.org/10.1080/10304312.2021.1937941.

Bourdieu, P. 1977 [1972]. *Outline of a Theory of Practice*. Translated by R. Nice. Cambridge: Cambridge University Press. https://doi.org/10.1017/CBO9780511812507.

Büchi, M., N. Festic and M. Latzer. 2018. 'How social wellbeing is affected by digital inequalities'. *International Journal of Communication* 12: 3686–706. https://ssrn.com/abstract=3871751.

de Certeau, M. 1984. *The Practice of Everyday Life*. Berkeley, CA: University of California Press.

Chadwick, A. 2007. 'Digital network repertoires and organisational hybridity', *Political Communication* 24(3): 283–301. https://doi.org/10.1080/10584600701471666.

Coe, C. 2013. *The Scattered Family: Parenting, African migrants, and global inequality*. Chicago, IL: University of Chicago Press.

Cozza, M., A. de Angeli and L. Tonolli. 2017. 'Ubiquitous technologies for older people', *Personal and Ubiquitous Computing* 21(3): 607–19. https://doi.org/10.1007/s00779-017-1003-7.

Czaja, S. J. and C. C. Lee. 2007. 'The impact of ageing on access to technology', *Universal Access in the Information Society* 5: 341–49. https://doi.org/10.1007/s10209-006-0060-x.

van Deursen, A., J. van Dijk and O. Peters. 2011. 'Rethinking internet skills: The contribution of gender, age, education, internet experience and hours online to medium-and content-related internet skills', *Poetics* 39(2): 125–44. https://doi.org/10.1016/j.poetic.2011.02.001.

Donner, J. 2015. *After Access: Inclusion, development and a more mobile internet*. Cambridge, MA: MIT Press.

Ebrey, J. 2016. 'The mundane and insignificant, the ordinary and the extraordinary: Understanding everyday participation and theories of everyday life', *Cultural Trends* 25(3): 158–68. https://doi.org/10.1080/09548963.2016.1204044.

Fernández-Ardèvol, M., K. Sawchuk and L. Grenier. 2017. 'Maintaining connections: Octo- and nonagenarians on digital "use and non-use"', *Nordicom Review* 38(1): 39–51. https://doi.org/10.1515/nor-2017-0396.

Friemel, T. N. 2016. 'The digital divide has grown old: Determinants of a digital divide among seniors', *New Media & Society* 18(2): 313–31. https://doi.org/10.1177/1461444814538648.

Gallistl, V., R. Rohner, A. Seifert and A. Wanka. 2020. 'Configuring the older non-user: Between research, policy and practice of digital exclusion', *Social Inclusion* 8(2): 233–43. https://doi.org/10.17645/si.v8i2.2607.

Haapio-Kirk, L. 2024. *Ageing with Smartphones in Japan: Reimagining care in a visual digital age*. London: UCL Press. https://doi.org/10.14324/111.9781787355767.

Hänninen, R., S. Taipale and R. Luostari. 2020. 'Exploring heterogeneous ICT use among older adults: The warm experts' perspective', *New Media & Society* 23(6): 1584–1601. https://doi.org/10.1177/146144482091735.

Hänninen, R., J. Karhinen, V. Korpela, L. Pajula, O. Pihlajamaa, M. Merisalo, O. Kuusisto, S. Taipale, J. Kääriäinen and T. A. Wilska. 2021a. *'Digiosallisuuden käsite ja keskeiset osa-alueet: Digiosallisuus Suomessa -hankkeen väliraportti'* ['The concept of digital inclusiveness and its key elements. An interim report on the "Digital Inclusiveness in Finland" project']. Publications of the Governments Analysis, Assessment and Research Activities 2021, 25. Accessed 7 December 2024. http://urn.fi/URN:ISBN:978-952-383-287-9.

Hänninen, R., S. Taipale and A. Korhonen. 2021b. 'Refamilisation in the broadband society – the effects of ICTs on family solidarity in Finland', *Journal of Family Studies* 27(2): 161–77. https://doi.org/10.1080/13229400.2018.1515101.

Hänninen, R., V. Korpela, L. Pajula and S. Taipale. 2022. 'Tuetun käytön merkitys informaalissa oppimisessa' ['The role of supported use in informal learning']. In *Ikääntyvät oppijat digitaalisessa mediayhteiskunnassa. Elinikäisen oppimisen mahdollisuudet* [Ageing learners in digital media society: Opportunities for lifelong learning], edited by E. Kärnä, P. Rasi, H. Vuojärvi, K. Pihlainen and K. Korjonen-Kuusipuro, 124–49. Helsinki: Gaudeamus.

Hänninen, R., V. Korpela and L. Pajula. 2023a. 'The paradoxes and pragmatics of digital leisure in later life', *Leisure Studies* 1–14. https://doi.org/10.1080/02614367.2023.2271194.

Hänninen, R., L. Pajula, V. Korpela and S. Taipale. 2023b. 'Individual and shared digital repertoires: Older adults managing digital services', *Information, Communication & Society* 26(3): 568–83. https://doi.org/10.1080/1369118X.2021.1954976.

Hargittai, E. 2002. 'Second-level digital divide: Differences in people's online skills', *First Monday* 7(4). https://doi.org/10.5210/fm.v7i4.942.
Hargittai, E., A. M. Piper and M. R. Morris. 2019. 'From internet access to internet skills: Digital inequality among older adults', *Universal Access in the Information Society* 18: 881–90. https://doi.org/10.1007/s10209-018-0617-5.
Hartman, M. 2008. 'Everyday life: Domesticating the invisible'. In *Innovating For and By Users*, edited by J. Pierson, E. Mante-meijer, E. Loos and B. Sapio, 3–12. Office for Official Publications of the European Communities.
Hasebrink, U. and J. Popp. 2006. 'Media repertoires as a result of selective media use: A conceptual approach to the analysis of patterns of exposure', *Communications* 31(3): 369–87. https://doi.org/10.1515/COMMUN.2006.023.
Hasebrink, U. and H. Domeyer. 2012. 'Media repertoires as patterns of behaviour and as meaningful practices: A multimethod approach to media use in converging media environments', *Participations: Journal of audience & reception studies* 9(2): 757–79.
Helsper, E. J. and A. J. van Deursen. 2017. 'Do the rich get digitally richer? Quantity and quality of support for digital engagement', *Information, Communication & Society* 20(5): 700–14. https://doi.org/10.1080/1369118X.2016.1203454.
Hill, R., L. R. Betts and S. E. Gardner. 2015. 'Older adults' experiences and perceptions of digital technology: (Dis)empowerment, wellbeing and inclusion', *Computers in Human Behavior* 48: 415–23. https://doi.org/10.1016/j.chb.2015.01.062.
Korpela, V., L. Pajula and R. Hänninen. 2024. 'Investigating the multifaceted role of warm experts in enhancing and hindering older adults' digital skills in Finland', *International Journal of Lifelong Education* 43(5): 509–22. https://doi.org/10.1080/02601370.2024.2353176.
Kuoppamäki, S., R. Hänninen and S.Taipale. 2022. 'Enhancing older adults' digital inclusion through social support: A qualitative interview study'. In *Vulnerable People and Digital Inclusion: Theoretical and applied perspectives*, edited by P. Tatsou, 211–30). Cham, Switzerland: Springer International Publishing. https://doi.org/10.1007/978-3-030-94122-2_11.
Lefebvre, H. 1947. *The Critique of Everyday Life*. London: Verso.
Lewis, T., P. Langdon and J. Clarkson. 2006. 'Investigating the role of experience in the use of consumer products'. In *Designing Accessible Technology*, edited by J. Clarkson, P. Langdon and P. Robinson, 189–98. Cham, Switzerland: Springer International Publishing. https://doi.org/10.1007/1-84628-365-5_19.
Livingstone, S., G. Mascheroni and M. Stoilova. 2023. 'The outcomes of gaining digital skills for young people's lives and wellbeing: A systematic evidence review', *New Media & Society* 25(5): 1176–1202. https://doi.org/10.1177/14614448211043189.
Loos, E. and L. Ivan. 2022. 'Not only people are getting old, but the new media are too: Technology generations and the changes in new media use', *New Media & Society*, 26(6): 3588–3613. https://doi.org/10.1177/14614448221101783.
Mare, A. 2016. 'Baba Jukwa and the digital repertoires of connective action in a "competitive authoritarian regime": The case of Zimbabwe'. In *Digital Activism in the Social Media Era*, edited by B. Mutsvairo, 45–68. Cham, Switzerland: Palgrave Macmillan. https://doi.org/10.1007/978-3-319-40949-8_3.
Mauss, M. 2006. *Techniques, Technology and Civilization*. New York, NY: Berghahn Books.
McLaughlin, F. 2014. 'Senegalese digital repertoires in superdiversity: A case study from Seneweb', *Discourse, Context & Media* 4: 29–37. https://doi.org/10.1016/j.dcm.2014.03.004.
Nimrod, G. 2017. 'Older audiences in the digital media environment', *Information, Communication & Society* 20(2): 233–49. https://doi.org/10.1080/1369118X.2016.1164740.
Nyman, J. 2021. 'The everyday life of security: Capturing space, practice and affect', *International Political Sociology* 15(3): 313–37. https://doi.org/10.1093/ips/olab005.
Olphert, W. and L. Damodaran. 2013. 'Older people and digital disengagement: A fourth digital divide?', *Gerontology* 59(6): 564–70. https://doi.org/10.1159/000353630.
Pajula, L., R. Hänninen, V. Korpela and S. Taipale. 2024. 'Exploring stakeholder perspectives on user involvement in designing digital public services for older adults', *First Monday* 29(9). https://doi.org/10.5210/fm.v29i9.13330.
P. de Paula Couto, M. C., J. Nikitin, S. Graf, H. H. Fung, T. M. Hess, S. Liou and K. Rothermund. 2023. 'Do we all perceive experiences of age discrimination in the same way? Cross-cultural differences in perceived age discrimination and its association with life satisfaction', *European Journal of Ageing* 20(43). https://doi.org/10.1007/s10433-023-00790-x.

Peek, S. T., K. G. Luijkx, M. D. Rijnaard, M. E. Nieboer, C. S. van der Voort, S. Aarts, J. van Hoof, H. J. Vrijhoef and E. J. Wouters. 2016. 'Older adults' reasons for using technology while ageing in place', *Gerontology* 62(2): 226–37. https://doi.org/10.1159/000430949.

Pérez-Amaral, T., A. Valarezo, R. López and T. Garín-Muñoz. 2021. 'Digital divides across consumers of internet services in Spain using panel data 2007–2019. Narrowing or not?'. *Telecommunications Policy* 45(2): 102093. https://doi.org/10.1016/j.telpol.2020.102093.

Pink, S. 2012. *Situating Everyday Life: Practices and places*. London: Sage Publications Ltd. https://doi.org/10.4135/9781446250679.

Quan-Haase, A., M. Guan Ying and B. Wellman. 2017. 'Connected seniors: How older adults in East York exchange social support online and offline', *Information, Communication & Society* 20(7): 967–83. https://doi.org/10.1080/1369118X.2017.1305428.

Rosenberg, D. and S. Taipale. 2022. 'Social and satisfied? Social uses of mobile phone and subjective wellbeing in later life', *Human Technology* 18(1): 45–65. https://doi.org/10.14254/1795-6889.2022.18-1.4.

Scheerder, A., A. van Deursen and J. van Dijk. 2017. 'Determinants of internet skills, uses and outcomes. A systematic review of the second- and third-level digital divide', *Telematics and Informatics* 34(8): 1607–24. https://doi.org/10.1016/j.tele.2017.07.007.

Seifert, A. and J. Rössel. 2019. 'Digital participation'. In *Encyclopedia of Gerontology and Population Ageing*, edited by D. Gu and M. Dupre, 1–5. Cham, Switzerland: Springer International Publishing. https://doi.org/10.1007/978-3-319-69892-2_1017-1.

Selwyn, N. 2004. 'Reconsidering political and popular understandings of the digital divide', *New Media & Society* 6(3): 341–62. https://doi.org/10.1177/1461444804042519.

Stevenson, I. 2013. 'Does technology have an impact on learning? A fuzzy set analysis of historical data on the role of digital repertoires in shaping the outcomes of classroom pedagogy', *Computers & Education* 69: 148–58. https://doi.org/10.1016/j.compedu.2013.07.010.

Swidler, A. 1986. 'Culture in action: Symbols and strategies', *American Sociological Review* 51(2): 273–86. https://doi.org/10.2307/2095521.

Tsatsou, P., S. Kurnia and J. Choudrie. 2018. 'Introduction'. In *Social Inclusion and Usability of ICT-enabled Services*, edited by J. Choudrie, S. Kurnia and P. Tsatsou, 11–18. New York, NY: Routledge.

Vehovar, V., P. Sicherl, T. Hüsing and V. Dolnicar. 2006. 'Methodological challenges of digital divide measurements', *The Information Society* 22(5): 279–90. https://doi.org/10.1080/01972240600904076.

2 From digital skills to digital repertoires: towards a more inclusive conceptualisation of technology use in an ageing society

Loredana Ivan

Introduction

Current literature on digital society often refers to digital skills as key in creating advantages and opportunities – or, conversely, disadvantages and inequalities (Hüsing et al. 2015; Kiss 2017). The concept of a 'digital divide' has been built on the idea that there are differences in usage and skills between certain groups of people, with those more affected by related vulnerability being women, older people and those with lower levels of education and/or lower incomes (van Deursen & van Dijk 2010; 2011). In such rhetoric, older people are normally portrayed as lacking digital skills or access to the latest communication technologies (see Ivan and Loos 2023).

To offer a deeper analysis of the 'digital skills' concept in the context of digitalisation today, this chapter argues instead for use of the term 'digital repertoire'. This concept, which describes the everyday practices of technology use, is more inclusive and brings nuance to the understanding of the way in which different social groups form part of the technological ecology (Edgerly et al. 2018; Peters et al. 2022). This in turn expands on the analysis offered by the dominant 'digital skills' concept by describing the relation between people and technology, predominant in today's global context of competition, fluid labour markets and digital transformations. The Covid-19 pandemic has also accentuated the prominence of the concept of 'digital skills' and blocked attempts to critique it; in the unprecedented situation created by the health crisis, digital prerequisites proved to be an important element in survival (Antonopoulou et al. 2021; Garcia et al. 2021).

Subsequently, the chapter draws attention to how the concept of digital skills has been measured. Traditionally self-assessment and performance tests, a metric that evaluates people's readiness for the future labour market, have been used. In this way, the term raises serious concerns about the normativity and linearity of the 'digital skills' concept, as well as its potential to reinforce a range of prejudices, including ageism, in describing the digital competencies of older adults. Furthermore, the concept of twenty-first-century digital skills is presented in the chapter as a non-linear and less reductionist way of measuring people's digital skills. It includes a broader perspective of the role of digital transformations in people's lives.

Finally, this chapter departs from a behaviourist approach ('more is better') to a functionalist perspective (based on the role played by different technologies in people's lives) by discussing the value of using 'digital repertoires'. As noted above, this concept serves to reveal the ways in which people navigate digital realities today, including the norms and conventions of digital interactions. The concept of digital repertoires allows not only for the investigation of individual digital use, but also for the co-use, supported use, non-use and discontinued use of digital technologies.

Investigating digital repertoires through ethnographic and narrative research (Hänninen et al. 2021a), as well as the use of longitudinal and cross-cultural data (see Ivan and Nimrod 2021), offers a focus on digital practices. This chapter also puts forward the argument that repertoires evolve through time, place and available technologies. It concludes by considering the role of digital repertoires for policy-making, providing an in-depth exploration of current digital realities.

Dominant role of digital skills in digital society discourse

The rise of the network society (Castells 2010 [1996]; 2014) accentuates the process of individuation and shapes a global culture organised around computer-mediated communication. The information age and the rapid integration of information and communication technologies (ICT) create an imperative to discuss and measure people's digital skills. As digital technology has become an integral part of both our personal and professional lives, having a transformative effect over the past 15 to

20 years, people necessarily have developed new skills. Moreover, public rhetoric emphasises that individuals should 'look for and develop' these new types of skills (Kiss 2017, 3) to be successful as workers and citizens.

Generally speaking, decision-makers and public bodies have imposed a hegemonic view on the concept of digital skills. For example, a report on digital skills in the EU labour market explicitly states that:

> Everybody will need to have at least basic digital skills to live, work, learn and participate in society, and a digitally skilled workforce will need to be present in the labour market to avoid skill gaps or skills mismatches (Kiss 2017, 3).

While digital technology is required in nearly every kind of job, the overall concern is that some people will be left behind, at risk of becoming marginalised or a burden to society (Hüsing et al. 2015). In addition, digital skills are perceived as important prerequisites from the point of view of competitiveness: societies, lifestyles and individuals who are better at acquiring advanced digital skills in response to the demands of the current digital transformations will secure their place at the top of global competition (Ovanessoff and Purdy 2011).

The overarching status of digital skills draws from the digitalisations of society in general, and of working life in particular. It forces a Cartesian representation of digital reality, in which the gap between 'haves' and 'have nots' is continuously highlighted (van Deursen and van Dijk 2011), compelling individuals to conform to a standardised list of skills (see, for example, ECDL Certification – an IT certification programme for basic computer skills recognised worldwide).[1] As a result, governments and international bodies urge that such skills are measured, showing differences between social groups (European Commission, *COM 2010/245*). The idea of 'digital skills' as a prerequisite for employment is illustrated by data showing that more than 90 per cent of current jobs require some level of digital skills (Feijao 2021). Even in the context of advanced rates of unemployment, numerous jobs in the ICT sector remained unfilled.

The Covid-19 pandemic accentuated the importance of digital skills (Antonopoulou et al. 2021; Manco-Chavez et al. 2020). The idea that we can no longer survive in contemporary society without proper digital competence (a set of digital skills that could be assessed, for example Calvani et al. 2008) is implicitly accepted by individuals and governments (Tsekeris 2019). Indeed, in certain circumstances digital literacy, which refers to a multiplicity of digital skills and knowledge of how to apply these skills in real-life situations (Ng 2012), proved to be a vital resource during the pandemic. For example, online education was crucial: educators, students and librarians had to learn new ways of transmitting information and interacting using internet-based tools (Manco-Chavez et al. 2020). Even the most critical voices on the role of new technologies in education had to submit to the new digital work environment and hone their skills accordingly. Again, the discussion remains in terms of those who succeeded in adequately acquiring the relevant skills and those who did not.

The Covid-19 pandemic also put pressure on 'vulnerable groups' (e.g. those with the lowest incomes or levels of education or the older members of society) to improve their digital skills – including those already acknowledged to have lower digital skills, such as older adults (Garcia et al. 2021). Studies suggest that ICTs can reduce the effects of social isolation (e.g. Buléon et al. 2022), providing healthcare at a distance, sustaining remote work and education (Al-Habaibeh et al. 2021) and also maintaining basic services (O'Sullivan et al. 2021), such as shopping and appointments (see, for example, Baker et al. 2018). However, these studies have mainly used self-assessment measures, in which people report the frequency of use, the type of use (for example, for work or leisure) and possible difficulties in using digital resources.

Digital skills are considered to be the main prerequisite of work adaptation. The development of digital skills in the pandemic scenario was focused on vulnerable social groups (Garcia et al. 2021). Raising the level of digital skills in such groups was presented as an emergency call for action. For example, a recent study on the computer skills of the EU population aged 55 years and above living in community dwellings called for intervention: in two-thirds of the EU countries 30 per cent of the population aged 55 and above have no computer skills. The results showed, unsurprisingly, that older adults in general have the lowest level of computer skills, something that has been labelled as a problem needing EU 'action and intervention' (Midão et al. 2020, 31). Note, however, that the study is based on SHARE data from 2015 (Survey of Health, Ageing and Retirement in Europe), in which people self-reported their computer skills on 5-point Likert scale (ranging from 1 (Excellent) to 5 (Poor), or 6 – I have never used a computer (spontaneous answer)).

This study opens the discussion on the challenges faced when using the concept of digital skills in a normative way. It is not only a matter of how we measure digital skills or computer skills, as in the example presented here (Midão et al. 2020), but also of what exactly we measure when assessing digital or computer skills. Challenges in operationalisation and in providing a standardised list of skills need to be assessed.

Challenges of measuring digital skills

As shown here, two methodologies prevail regarding the measurement of digital skills (Allmann and Blank 2021): self-assessment measures and performance tests. Self-assessment approaches either ask respondents to evaluate their skills, as in the example above, or to answer whether they can perform certain tasks such as using a browser, shopping online or designing a website. An obvious limitation of self-rating digital skills is the lack of external validity and the fact that people have limited capacities to evaluate their own skills. There are certain groups of people who have particular systematic bias when they self-rate digital skills (Hargittai and Shafer 2006; van Deursen and van Dijk 2010). We know, for example, that older people tend to underestimate their digital competence, whereas for younger people the effect is the reverse: they tend to overestimate their digital skills (Hargittai et al. 2019). Furthermore, self-evaluation of digital skills occurs within a social context that should not be overlooked. Internalised stereotypes regarding a respondents' social group could play a role in shaping a response. This can be the case for older women, for instance, who tend to underestimate their digital skills compared to older men, following stereotypical gender roles that associate technology with masculinity (Ivan and Schiau 2016). In addition, the fact that advanced digital skills are presented as socially desirable is an important contextual factor when individuals self-report their skills (Zimbardo 1995).

Performance tests are more accurate measures of evaluating digital skills by asking participants to perform certain tasks while observed by researchers (van Deursen and van Dijk 2010). Besides the time and costs, a performance test is a top-down approach, starting from what is expected from people in terms of digital skills – an etic perspective – of what somebody should do to be labelled as 'digitally competent'. However, this gives us no information on the relevance of each task for people in their individual lives. Recently, more in-depth methodologies have been used in an attempt to construct an emic perspective on digital skills (Allmann and Blank 2021), based on ethnographic techniques such as observations and interviews. Even so, however, the way in which digital skills have been operationalised does not encourage the use of in-depth methodologies.

Traditionally, digital skills have been related to the technical aspects of the medium of communication (van Deursen et al. 2016; van Deursen and van Dijk 2011); they are generally defined as 'the capacity to respond pragmatically and intuitively to challenges and opportunities in a manner that exploits the internet's potential' (DiMaggio et al. 2004,

378). More recently, digital skills are defined in terms of effects – by focusing on positive outcomes, for example the ability to solve problems and to make use of online resources. Digital skills are considered 'the primary requirement for conducting capital-enhancing activities online, for obtaining positive outcomes from internet use' (van Deursen and Mossberger 2018, 123).

The concept of digital literacy (Glister 1998) changes the focus of digital skills from the ability to handle technology (for example, using a browser, creating a file and installing an application) to mastering ideas and addressing problems with the use of digital resources. Digital literacy then encompasses cognitive and social-emotional skills to help in problem-solving (Eshet-Alkalai 2004). Digital literacy is seen not only as a sum of digital skills but also as a mind-set (van Laar et al. 2017) that enables individuals 'automatically' to solve different issues in the digital environment, effectively and effortlessly using their technical, cognitive and socio-emotional resources (see Ng 2012).

The original context of the concept of digital skills was to investigate people's readiness for the future labour market. Following the labour-market-oriented definitions of the term contributes to its normativity and certain ageist characteristics of the concept, particularly when applying it to the theory of age-based digital divides.

When operationalising digital skills, the concept accounts for technical competence in communications (see van Deursen and van Dijk 2010), ranging from content-related skills (e.g. sharing and reacting to content) to information communication (e.g. using a digital environment to find relevant things) and to creative and strategic skills (e.g. using a digital environment for self-presentation). More specifically, lists of ICT abilities used for work, leisure, learning and communication have been deployed for operationalisation of the digital skills concept. For example, the European Parliament (2006) refers to

> the use of computers to retrieve, access, store, produce, present and exchange information, and to communicate and participate in collaborative networks via the internet (394).

The Organisation for Economic Co-operation and Development (OECD) launched a large programme for testing adults on different skills, including problem-solving in technology-rich environments.[2] In so doing it listed four layers of digital skills: 1) foundational skills, including emotional and social skills enabling the use of digital technologies; 2) complementary skills, such as information processing, problem-solving and communication in the digital environment; 3) generic skills related to the use of digital resources for professional purposes, such as handling commonly used software; 4) special skills, named e-skills by some authors (see Fernández-Sanz et al. 2017). These skills are required for professionals in the ICT sectors, for example in programming or app development.

The operationalisation of digital skills by international bodies such as the European Union and OECD underpins the idea of employability by focusing on the digital skills people would need to be successful in the labour market in the so-called information society. Nevertheless, it is worth asking: who decides what digital skills are needed for a certain period or setting? We know little about the process of decision-making: how such skills are selected, valued and considered desirable over time. Indeed, international bodies rely on experts' views (Donoso et al. 2020; van Laar et al. 2022) and estimations about how the labour market might evolve in the future, with some studies finding that desirable digital skills might not correspond to managers' expectations (see Prezioso et al. 2021). Whereas other studies highlight an unbalanced focus on certain skills – for example, prioritising information- and communication-related digital skills while ignoring other aspects such as strategic use and content creation (Iordache et al. 2017).

A recent and broader model to operationalise digital skills has been developed by van Laar et al. (2017). This model integrates the idea of twenty-first-century skills with the concept of digital skills, resulting in a hybrid term: 'twenty-first-century digital skills'. Through a systematic literature review, the researchers identified the core skills needed in a digital environment, one not restricted to the technical use of different digital resources and in line with twenty-first-century skills (as provided by the *Assessment and Teaching of 21st Century Skills* – Binkley et al. 2012). The conceptualisation of twenty-first-century digital skills comprises seven core dimensions: technical, information management, communication, collaboration, creativity, critical thinking and problem-solving. They also included five contextual dimensions; ethical awareness, cultural awareness, flexibility, self-direction and lifelong learning, as presented briefly in Figure 2.1.

The conceptual model of 'twenty-first-century digital skills' brings some nuance to the normativity of describing digital skills; it is also more context-sensitive than previous attempts have been. In addition, the model has been tested on a relatively large sample of employees (N=1.222) working in the creative industries (van Laar et al. 2019), proving that problem-solving was a key component of twenty-first-century digital skills

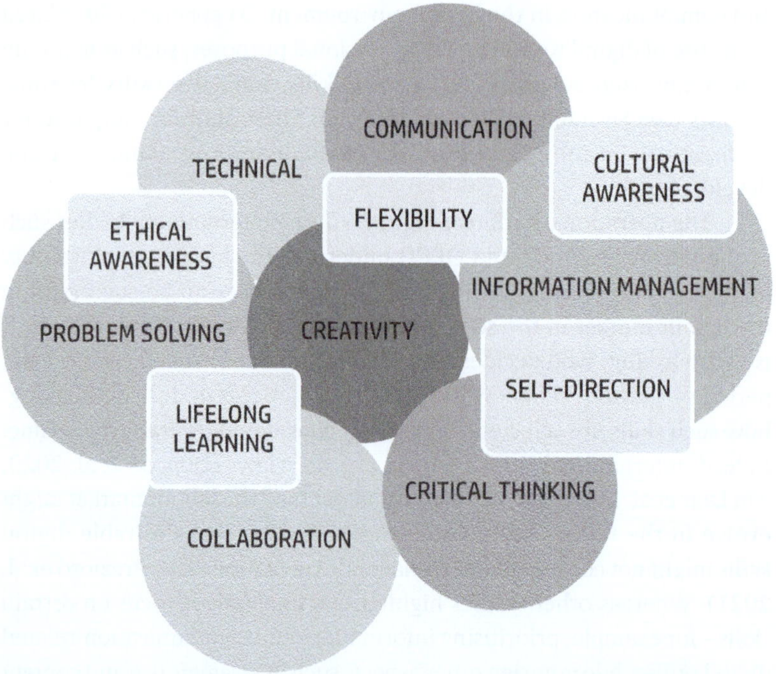

Figure 2.1 Operationalisation of twenty-first-century digital skills, adapted from van Laar et al. 2017. Image by the author.

– and also, more importantly, that the listed skills build on each other sequentially. As previous attempts treated the listed skills independently, current research on digital skills acknowledges the fact that the measured skills could be simultaneous or sequential but are not necessarily linear (Allmann and Blank 2021).

Nonetheless, the digital skills concept is deterministic in nature, forcing us to think in terms of hierarchies, privileging the 'advanced' over the 'laggards'. It is a concept largely used to assess individuals' success in the labour market and their potential employability in a society marked by rapid digital transformation. At the general level, the way it has been operationalised and measured responds to the idea of the global innovation competition. At the individual level, the concept of digital skills starts from the premise that 'more is better' and ignores the meanings people invest in digital technologies. Below I argue further for digital repertoires as an alternative and less deterministic concept that could be employed to understand the role of digital technologies in everyday life, and so go beyond people's capital in the labour market.

Moving from digital skills to digital repertoires

In recent years, the concept of digital repertoires has been employed to capture digital realities (Edgerly et al. 2018; Lee and Yang 2014; Peters et al. 2022). There is a shift from seeing technology use as a set of discrete behaviours and skills, aggregated at the individual level to place a person on a hierarchy, to everyday practices and mechanisms through which people navigate the technological ecology of everyday life. With digital repertoires, the focus is on the meanings of using different technologies and the significance that people invest in their digital behaviours and others' digital practices.

The repertoires are meaningful and fulfil their functional adaptation role: to meet people's needs in everyday situations (Schwarzenegger 2020). Beyond discrete digital behaviours, digital repertoires encompass a continuum of contextualised digital practices. This presents a shift from an etic to an emic perspective in understanding the adaptive role that digital tools play in people's lives. To give an example, when interviewing older Romanian women in the 'Grannies on the Net' project (Ivan and Nimrod 2021; Ivan and Mocanu 2020), we seldom found women who were not using Google applications to search for different information that they might have needed in daily routines (for example, the service hours of certain institutions). This would have counted as a low level of digital skill – mainly the ability to solve problems by using digital technology.

During interviews with some of these women, I would ask exactly how they would proceed when they sought to obtain particular information, exploring such practices 'from the inside'. It was a revelation that some of them used Google Voice Search to find information, even though they did not name the functionality as such. The conceptual strength of digital repertoire derives from the fact that they offer a holistic view of different practises that co-exist, in the service of a desired end-goal (Peters and Schrøder 2018; Peters et al. 2022). The concept also offers insight into the possibilities people have of navigating through different media spaces, and of using interrelated modes of different technological affordances (Taipale et al. 2021).

Indeed, the digital repertoire concept shares a connotation with the meaning of media repertoire – a term that has been used in the literature to explore media use (see Gallistl and Nimrod 2020; Hasebrink and Popp 2006; Hasebrink and Domeyer 2012; Hänninen et al. 2021a; Olsson et al. 2019). Starting from media ecology theory, introduced by McLuhan in 1964 and later developed by Postman in 1969 (Postman 2000), media repertoire investigates the interrelation between technology, the communication medium and the social environment. Media repertoires thus describe the entire media landscape of a person (see Gallistl and Nimrod 2020) and the contexts in which this could be beneficial.

The fact that some media uses could be more beneficial than others, for example in predicting wellbeing or reducing loneliness (Liu et al. 2019), could explain the preference for such media and their functional role. I agree with Nimrod (2017), who argues that current studies on media repertoires rarely offer a holistic and emic perspective. Such studies (see Boczkowski et al. 2018; Choi 2016) instead explore how the discrete use of different media are associated with certain positive or negative outputs in people's lives. In the case of older adults, for example, studies show that greater use of social media is accompanied by higher life satisfaction (e.g. Gaia et al. 2021); others, however, have produced contradictory results (e.g. Tomini et al. 2016; Zhao 2021).

In more nuanced studies (see Gallistl and Nimrod 2019; Nimrod and Ivan 2019), social media use and older adults' subjective wellbeing have been linked, revealing the functional role of different media. Some media have been perceived by people as 'ways of wasting time', whereas others are perceived as 'ways of filling the time' (Nimrod and Ivan 2019,16). When moving from a behaviourist ('the more the better') to a functionalist perspective, we incorporate the meaning of the concept of repertoire and depart from the deterministic view that is criticised here regarding the digital skills concept. Research on digital repertoires reveals media use in social context, in connection with people's lifestyles and the practical meanings of different social interactions (Schwarzenegger 2020).

The role of digital repertoires in describing digital practices

Although media and digital repertoires encompass similar connotations, digital repertoires offer a broader view of technology use in everyday life, transcending media use to cover internet-based devices, digital applications and services. Digital repertoires refer to media content and include affordability and intersectionality of the media content, as well as how such content is used in people's daily routine. However, 'digital skills' and 'digital repertoires' are intersecting concepts, referring both to people's technology appropriation and technology domestication. The fact that digital technologies are omnipresent in people's lives requires certain skills (competence) and functionalities associated to one's needs and daily routine. We can therefore understand that the two concepts form part of the same rhetoric on the ubiquitousness of digital influence in our lives.

The digital repertoire is not only a broader concept than media repertoire; it also offers a way of looking at the use of technology to create social accountability about the ways in which people navigate through today's digital realities. I use here the term of 'accountability' suggested by Harvey Sacks (1992) to describe the subtle ways in which our behaviours convey norms and conventions of everyday interactions. For example, in the case of older adults, as Hänninen et al. (2021b) have suggested, investigating people's digital repertoires would mean also discussing the co-use, supported use (as the role of the warm experts) and the non-use or discontinued use of digital technologies. Research often focuses on older adults' reluctance to adopt new technologies, or their 'lagging' ability to use different devices, or the challenges they face on being immersed in the new digital sphere. This is evident in the use of the concepts of 'digital immigrants' as opposed to 'digital natives' (Prensky 2001). We want better to understand the digital ecologies created by people, the environment and the technologies they are using. By doing so, as shown in a recent article (Loos and Ivan 2022), we aim better to comprehend the relation between age and technology use by underlining the fact that it is not only people who are getting old; digital technologies are too.

In addition to ethnographic and interview-based research methods (Hänninen et al. 2021a), the use of longitudinal data can be equally beneficial in the study of digital repertoires, showing how practices change and are maintained over time. Such studies highlight changes in the meanings people ascribe to different digital practices over time. Indeed, the concept of repertoire incorporates the idea of dynamics (Tilly 2006), as repertoires evolve 'with time, place and available technologies' and differ according to context (Khazraee and Losey 2016, 41). From this point of view, longitudinal and cross-cultural data (both quantitative and qualitative) may provide a better approach to the investigation of digital repertoires than cross-sectional and culturally homogeneous data.

A more nuanced way to approach digital repertoires would be an investigation of how daily life is reflected and reconstructed through the use of digital technology. The concept encompasses socio-digitisation (Latham and Sassen 2005, 3), the process by which daily life reflects and is reflected through technology use. By studying the ways in which digital technologies are used in everyday contexts, we have access to people's micro-social universe. For example, in previous work (see Ivan and Nimrod 2021), I have offered insight into family conflicts by investigating technology use in intergenerational settings, using narratives from grandmothers in different countries. In doing so, we shed light on the role of digital technologies in generating, maintaining and solving conflicts.

In other work, the routines of daily life offer a reflection of the digital spectrum, providing another way of approaching the concept of digital repertoires. For example, in one of our studies (Ivan and Mocanu 2020), we explored how a group of older people talk about how they have incorporated different digital technologies into their lives. To trigger their reflections, we used video material featuring a short artistic movie about an elderly couple struggling to install an internet-based application. We noticed that people tended to construct meanings for the different digital tools they have appropriated by associating them with their social environment and different situations in life. Some would say, for example, that receiving a phone specially designed for older people from their adult children was somehow humiliating, or at least uncomfortable. They thus ascribed a negative meaning to that type of device as an object that marked the loss of their agency with technology.

Moreover, this approach enabled us to talk about people's digital repertoires in general, as well as about specific digital repertoires – for example, blogging repertoires, social network sites (SNS) repertoires and instant messaging repertoires (see Khazraee and Losey 2016). People's digital repertoires became more nuanced with rapid change and diversified technology. The SNS repertoires, for example, do not only incorporate 'skills' to handle different options offered by SNSs, but also come together with a kind of tacit knowledge (Polanyi and Sen 2009; Stevenson 2013). Such an implicit recognition of norms and regularities would enable people not only to explore different functionalities of such digital platforms, but also to proceed in the 'proper way'.

A good example of how SNS repertoires differentiate between generations, creating irreconcilable situations, appears in an interview I recently conducted with a woman of 65 in Romania (Mandache and Ivan 2022). During the Covid-19 pandemic, she loved using Facebook to communicate. Every morning, she would post a 'good morning' message on her page, feeling connected to the world in this way. When her son asked her to stop because she was embarrassing him, she felt really confused. Repertoires can certainly be complementary, but they may also be conflictual. Digital repertoires mirror people's micro-sociality (in this instance, for example, the conflicts between generations). They therefore become a valuable tool to investigate some sensitive issues, such as people's intergenerational expectations or conflictual relationships.

In addition, the concept of digital repertoires offers a more useful, less prejudiced conceptual framework for studying technology use later in life. I agree with Taipale et al. (2017,4), that age-related digital gaps are specific to certain periods and will eventually disappear, and that the 'digital skills' deterministic approach seems unable to capture the integration of technology in older adults' everyday lives.

There are two principal reasons for this. Firstly, the concept itself is related to the work environment and to people's success in the labour market, something that many older adults have already left. The whole idea of improving people's skills serves the purposes of readiness for the new digitalised global economy and the increased level of competitiveness between economies from different parts of the world. The digital life outside the working sphere is difficult to capture through the concept of 'digital skills'.

Secondly, when measuring 'digital skills' older adults will always be the later adopters, the least advanced and in some ways 'behind' the younger groups. This serves to create and reinforce a stereotypical representation of older adults as being unable to use digital technology, or at least as being 'not that good at it'. In a previous article (Ivan and Cutler 2021) we showed how powerful such ageist views are and how they activate the 'vicious cycle of ageism' in which people internalised such representations, maintaining themselves in the so-called 'grey digital divide' (Friemel 2016).

As noted in this chapter, older adults tend to underrate their digital skills, whereas younger adults may often overrate theirs (see e.g. Barrie et al. 2021; Černochová et al. 2020; Hargittai et al. 2019; Ivan and Schiau 2016; Maderick et al. 2016). We also showed how when people underrate their abilities, a process of self-fulfilling prophecy may occur – such a vicious cycle is difficult to break. As we have observed elsewhere, the 'stereotype' issue plays a role in older people's underperformance when using digital technologies (Ivan and Cutler 2021).

Conclusion: from more inclusive concepts to more inclusive societies

The current chapter argues that 'digital skills' is a concept that dominates the rhetoric of public bodies in the discussion about contemporary digitalised society. The normativity and linearity of such rhetoric may have negative consequences in reinforcing prejudices and social cleavages between those who have the proper digital skills and those who do not, creating arbitrary hierarchies and ultimately a divided society. However, digital repertoire could be a more inclusive concept when compared to 'digital skills'. Moreover, the use of digital skills is related to people's employability and competitiveness in the labour market. The study of digital technology use should extend beyond the traditional work sphere, particularly in reference to older people. In addition, the decision regarding what digital skills are relevant for success in the labour market might also be arbitrary, or at least problematic.

The critical reaction of policy-makers to a more nuanced conceptual framework and the study of digital repertoires could be related to the fact that policy development is often based on the measuring of several indicators which are easily quantified. This is the main limitation of the practical use of the digital repertoire concept instead of digital skills. Nevertheless, although the digital repertoire concept does not focus on the quantification of digital behaviours, it does provide a rather more in-depth exploration of current digital realities. Here I have offered a reaction to such criticism by explaining that longitudinal data (both quantitative and qualitative), as well as data coming from different cultural groups, are a novel way of capturing the dynamics and multiplicity of digital practices – the essence of the 'digital repertoire' concept. However, longitudinal data are not always available and are not even country-inclusive: data are mostly collected in Europe and North America, and longitudinal studies are costly and difficult to conduct.

The ACT longitudinal study (2016–2021, see Loos et al. 2018; 2020), in which we investigated digital practices among older people across seven countries (Austria, Canada, Finland, Israel, the Netherlands, Romania and Spain), gathering panel data in three waves (2016, 2018 and 2020), is rather an exception. Although there are longitudinal and cross-national studies collecting data about digital skills (see, for example, OECD 2016 – PIAAC longitudinal study), they are normally focused on young adults, and they target skills that could be converted into valuable outputs on the labour market. In addition, little is known about the contexts in which such skills have been acquired and adjusted, or indeed about the contexts in which they are used. Moreover, people of all ages use their digital repertoires in contexts that transcend the work environment. They may be used to connect with others and reduce loneliness, for example, or to look for a partner or to stay in touch with topics related to hobbies, values and interests. I believe that 'digital repertoires' would be equally relevant to understanding the wellbeing of these users.

After all, a concept is not only a theoretical construct; it is also a way of seeing the world through certain 'lenses'. A reliance on the concept of digital skills envisages a more deterministic world in which success is rated by the way we adapt faster to predefined digital realities, whereas digital repertoires give room for individual choices and tacit digital knowledge, for diversity and subjective meanings. Furthermore, digital repertoires are not obliged to be consistent or free from contradictions. For example, an older adult can be an avid photographer, experienced in using sophisticated equipment and software, but at the same time be less familiar with email (because it involves typing) or with using digital services.

People can also change or become more sophisticated in their repertoires. Understanding and appreciating such change would mean (re-)considering the relevance of different digital practices in their lives. In failing to understand people's perspective towards what they are doing when using digital devices, we will eventually employ a 'technology-centred' lens, with possible negative consequences. The concept of technostress, for example (Ayyagari et al. 2011), coined by the clinical psychologist Crain Brod (1984), describes a modern disease resulting from an inability to cope with the proliferation of information and communication technologies (Ayyagari et al. 2011). Technostress phenomena have now become the subject of widespread research (Nimrod 2022; Tarafdar et al. 2019; 2020), revealing the discomfort and unease people may experience when it is imperative to engage constantly with new (digital) technologies and become more advanced or skilled. Those who fail to keep up with increasingly ubiquitous technological devices and do not learn how to handle the latest technology advancements will eventually be left behind.

This underscores the importance of critiquing digital skills as a less inclusive concept that stresses technology use and creates a binary division between 'successful' and 'unsuccessful' individuals. Instead, I argue that the idea of digital repertoires helps us to implement a person-environment fit model, providing a more contextual and nuanced understanding of our interaction with a range of digital technologies.

Acknowledgements

This project was funded by the Executive Agency for Higher Education, Research, Development and Innovation Funding (UEFISCDI), *UnLoc project*, Co-fund Urban Transformation Capacities (ENUTC), Grant Agreement No. 101003758.

Notes

1 See https://icdleurope.org/.
2 Programme for the International Assessment of Adult Competencies, PIAAC.

References

Al-Habaibeh, A., M. Watkins, K. Waried and M. B. Javareshk. 2021. 'Challenges and opportunities of remotely working from home during the Covid-19 pandemic', *Global Transitions* 3: 99–108. https://doi.org/10.1016/j.glt.2021.11.001.

Allmann, K. and G. Blank. 2021. 'Rethinking digital skills in the era of compulsory computing Methods, measurement, policy and theory', *Information, Communication & Society* 24(5): 633–48. https://doi.org/10.1080/1369118X.2021.1874475.

Antonopoulou, H., C. Halkiopoulos, O. Barlou and G. N Beligiannis. 2021. 'Transformational leadership and digital skills in higher education institutes: During the Covid-19 pandemic', *Emerging Science Journal 5*(1): 1–15. https://doi.org/10.28991/esj-2021-01252.

Ayyagari, R., V. Grover and R. Purvis. 2011. 'Technostress: Technological antecedents and implications', *MIS Quarterly* 35(4): 831–58. https://doi.org/10.2307/41409963.

Baker, S., J. Warburton, J. Waycott, F. Batchelor, T. Hoang, B. Dow and F. Vetere. 2018. 'Combatting social isolation and increasing social participation of older adults through the use of technology: A systematic review of existing evidence', *Australasian Journal of Ageing* 37(3): 184–93. https://doi.org/10.1111/ajag.12572.

Barrie, H., T. La Rose, B. Detlor, H. Julien and A. Serenko. 2021. 'Because I'm old': The role of ageism in older adults' experiences of digital literacy training in public libraries', *Journal of Technology in Human Services* 39(4): 379–404. https://doi.org/10.1080/15228835.2021.1962477.

Binkley, M., O. Erstad, J. Herman, S. Raizen, M. Ripley, M. Miller-Ricci and M. Rumble. 2012. 'Defining twenty-first-century skills'. In R. Pingge (Ed.), *Assessment and Teaching of 21st-century Skills*, edited by R. Pingge, 17–66.Dordrecht: Springer International Publications.

Boczkowski, P. J., M. Matassi and E. Mitchelstein. 2018. 'How young users deal with multiple platforms: The role of meaning-making in social media repertoires', *Journal of Computer-Mediated Communication*, 23(5): 245–59. https://doi.org/10.1093/jcmc/zmy012.

Brod, C. 1984. *Technostress: The human cost of the computer revolution*. Redwood City, CA: Addison-Wesley.

Buléon, C., J. Caton, Y. S. Park, S. Eller, M. Buyck, S. Kardong-Edgren and J. C. Palaganas. 2022. 'The state of distance healthcare simulation during the COVID-19 pandemic: Results of an international survey', *Advances in Simulation* 7(1): 1–11. https://doi.org/10.1186/s41077-022-00202-7.

Calvani, A., A. Cartelli, A. Fini and M. Ranieri. 2008. 'Models and instruments for assessing digital competence at school', *Journal of e-Learning and Knowledge Society, 4*(3): 183–93. https://doi.org/10.20368/1971-8829/288.

Castells, M. 2010 [1996]. *The Rise of the Network Society: The Information Age: Economy, society and culture*, vol. 1, second edition. Chichester, West Sussex, UK: Wiley-Blackwell.

Castells, M. 2014. 'The impact of the internet on society: A global perspective', *Change* 19: 127–48. https://www.bbvaopenmind.com/en/articles/the-impact-of-the-internet-on-society-a-global-perspective/.

Černochová, M., H. Voňková, J. Štípek and P. Černá. 2020. 'How do learners perceive and evaluate their digital skills?', *International Journal of Smart Education and Urban Society* 9(1): 34–47.

Choi, J. 2016. 'Why do people use news differently on SNSs? An investigation of the role of motivations, media repertoires and technology cluster on citizens' news-related activities', *Computers in Human Behavior* 54: 249–56. https://doi.org/10.1016/j.chb.2015.08.006.

van Deursen, A. and J. A. van Dijk. 2010. 'Measuring internet skills', *International Journal of Human–Computer Interaction* 26(10): 891–916.

van Deursen, A. and J. A. van Dijk. 2011. 'Internet skills and the digital divide', *New Media & Society* 13(6): 893–911. https://doi.org/10.1177/1461444810386774.

van Deursen, A. J., E. J. Helsper and R. Eynon. 2016. 'Development and validation of the Internet Skills Scale (ISS)', *Information, Communication & Society* 19(6): 804–23. https://doi.org/10.1080/1369118X.2015.1078834.
van Deursen, A. and K. Mossberger. 2018. 'Anything for anyone? A new digital divide in Internet-of-things skills', *Policy & Internet* 10(2): 122–40. https://doi.org/10.1002/poi3.171.
DiMaggio, P., E. Hargittai, C. Celeste and S. Shafer. 2004. 'Digital inequality: From unequal access to differentiated use', *Social Inequality* 1: 355–400.
Donoso, V., J. Pyżalski, N. Walter, N. Retzmann, A. Iwanicka, L. d'Haenens and K. Bartkowiak. 2020. *Report on Interviews with Experts on Digital Skills in Schools and the Labour Market.* KU Leuven: ySKILLS, https://doi.org/10.5281/zenodo.4274613.
Edgerly, S., E. K. Vraga, L. Bode, K. Thorson and E. Thorson. 2018. 'New media, a new relationship to participation? A closer look at youth news repertoires and political participation', *Journalism & Mass Communication Quarterly* 95(1): 192–212. https://doi.org/10.1177/1077699017706928.
Eshet-Alkalai, Y. 2004. 'Digital literacy: A conceptual framework for survival skills in the digital era', *Journal of Educational Multimedia and Hypermedia* 13(1): 93–106.
European Commission (2010). *COM 2010/ 245. A digital agenda for Europe.* Accessed 15 December 2024. EUR-Lex - 52010DC0245 - EN - EUR-Lex (europa.eu).
European Parliament. 2006. *Recommendation of the European Parliament and the Council of 18 December 2006 on Key Competencies for Lifelong Learning (2006/962/EC).* Official Journal of the European Union. Accessed 15 December 2024. https://eur-lex.europa.eu/legal-content/EN/TXT/PDF/?uri=CELEX:32006H0962.
Feijao, C., I. Flanagan, C. van Stolk and S. Gunashekar. 2021. *The Global Digital Skills Gap: Current trends and future directions.* https://www.rand.org/pubs/research_reports/RRA1533-1.html.
Fernández-Sanz, L., J. Gómez-Pérez and A. Castillo-Martínez. 2017. 'E-Skills Match: A framework for mapping and integrating the main skills, knowledge and competence standards, and models for ICT occupations', *Computer Standards & Interfaces* 51: 30–42. https://doi.org/10.1016/j.csi.2016.11.004.
Friemel, T. N. 2016. 'The digital divide has grown old: Determinants of a digital divide among seniors', *New Media & Society* 18(2): 313–31. https://doi.org/10.1177/1461444814538648.
Gaia, A., E. Sala and G. Cerati. 2021. 'Social networking sites use and life satisfaction. A quantitative study on older people living in Europe', *European Societies* 23(1): 98–118. https://doi.org/10.1080/14616696.2020.1762910.
Gallistl, V. and G. Nimrod. 2020. 'Media-based leisure and wellbeing: A study of older internet users', *Leisure Studies* 39(2): 251–65. https://doi.org/10.1080/02614367.2019.1694568.
Gallistl, V. and G. Nimrod. 2019. 'Online leisure and well-being in later life'. In *Perspectives on Human-Computer Interaction Research with Older People*, edited by S. Savage, 139–54. Cham, Switzerland: Springer International Publishing. https://doi.org/10.1007/978-3-030-06076-3_9.
Garcia, K. R., L. Rodrigues, L. Pereira, G. Busse, M. Irbe, M. Almada and E. Costa. 2021. 'Improving the digital skills of older adults in a COVID-19 pandemic environment', *Educational Gerontology* 47(5): 196–206. https://doi.org/10.1080/03601277.2021.1905216.
Glister, P. 1998. *Digital Literacy*. New York: Wiley.
Hänninen, R., L. Pajula, V. Korpela and S.Taipale. 2021a. 'Individual and shared digital repertoires: Older adults managing digital services', *Information, Communication & Society* 26(3): 568–83. https://doi.org/10.1080/1369118X.2021.1954976.
Hänninen, R., S. Taipale and A. Korhonen. 2021b. 'Re-familiarisation in the broadband society: The effects of ICTs on family solidarity in Finland', *Journal of Family Studies* 27(2): 161–77. https://doi.org/10.1080/13229400.2018.1515101,
Hargittai, E., A. M. Piper and M. R. Morris. 2019. 'From internet access to internet skills: Digital inequality among older adults', *Universal Access in the Information Society* 18(4): 881–90. https://doi.org/10.1007/s10209-018-0617-5.
Hargittai, E. and S. Shafer. 2006. 'Differences in actual and perceived online skills: The role of gender', *Social Science Quarterly* 87(2): 432–48. https://doi.org/10.1111/j.1540-6237.2006.00389.x.
Hasebrink, U. and J. Popp. 2006. 'Media repertoires as a result of selective media use. A conceptual approach to the analysis of patterns of exposure', *Communications* 31(2): 369–87. https://doi.org/10.1515/COMMUN.2006.023.

Hasebrink, U. and H. Domeyer. 2012. 'Media repertoires as patterns of behaviour and as meaningful practices: A multimethod approach to media use in converging media environments', *Participations* 9(2): 757–79.

Hüsing, T., W. B. Korte and E. Dashja. 2015. 'E-skills in Europe: Trends and forecasts for the European ICT professional and digital leadership'. *Labour Markets* (2015–2020), European Central Bank. Empirica Working Papers. https://eufordigital.eu/wp-content/uploads/2019/10/e-Skills-in-Europe_Trends-and-Forecasts.pdf.

Iordache, C., I. Mariën and D. Baelden. 2017. 'Developing digital skills and competencies: A quick scan analysis of 13 digital literacy models', *Italian Journal of Sociology of Education* 9(1): 6–30.

Ivan, L. and I. Schiau. 2016. 'Experiencing computer anxiety later in life: The role of stereotype threat'. In *Human Aspects of IT for the Aged Population. Design for aging*, edited by Z. Zhou and G. Salvendy, 339–49. Cham, Switzerland: Springer International Publishing. https://doi.org/10.1007/978-3-319-39943-0_33.

Ivan, L. and F. Mocanu. 2020. 'Older people and technology use: The importance of using video stimuli in group discussions'. In *Human Aspects of IT for the Aged Population. Technologies, Design, and User Experience*, edited by Q. Gao and J. Zhou, 147–63. Cham, Switzerland: Springer International Publishing. https://doi.org/10.1007/978-3-030-50252-2_12.

Ivan, L. and S. J. Cutler. 2021. 'Ageism and technology: The role of internalized stereotypes', *Special Issue of the University of Toronto Quarterly on Ageism* 90(2): 127–39. https://doi.org/10.3138/utq.90.2.05.

Ivan, L. and G. Nimrod. 2021. 'Family conflicts and technology use: The voices of grandmothers', *Family Relations* 70(1): 104–19. https://doi.org/10.1111/fare.12530.

Ivan, L. and E. Loos. 2023. 'The marketing of technology products for older people'. In *Digital Ageism: How it operates and approaches to tackling it*, edited by A. Rosales, M. Fernández-Ardèvol and J. Svensson, 88–115. London: Routledge. https://doi.org/10.4324/9781003323686.

Khazraee, E. and J. Losey. 2016. 'Evolving repertoires: Digital media use in contentious politics', *Communication and the Public* 1): 39–55. https://doi.org/10.1177/2057047315625076.

Kiss, M. 2017. *Digital skills in the EU labour market*. European Parliamentary Research Service, ERPS. Accessed 15 December 2024. https://www.europarl.europa.eu/RegData/etudes/IDAN/2017/595889/EPRS_IDA(2017)595889_EN.pdf.

van Laar, E., A. J. van Deursen, J. A. van Dijk and J. De Haan. 2017. 'The relation between 21st-century skills and digital skills: A systematic literature review', *Computers in Human Behavior* 72: 577–88. https://doi.org/10.1016/j.chb.2017.03.010.

van Laar, E., A. J. van Deursen, J. A. van Dijk and J. De Haan. 2019. 'Determinants of 21st-century digital skills: A large-scale survey among working professionals', *Computers in Human Behavior* 100: 93–104. https://doi.org/10.1016/j.chb.2019.06.017.

van Laar, E., A. J. van Deursen and J. A. van Dijk. 2022. 'Developing policy aimed at 21st-century digital skills for the creative industries: An interview study with founders and managing directors', *Journal of Education and Work* 35(2): 195–209. https://doi.org/10.1080/13639080.2022.2036710.

Latham, R. and S. Sassen. 2005. 'Digital formations: Constructing an object of study'. In *Digital Formations IT and New Architectures in the Global Realm*, edited by R. Latham and S. Sassen, 1–34. Princeton, NJ: Princeton University Press.

Lee, H. and J. Yang. 2014. 'Political knowledge gaps among news consumers with different news media repertoires across multiple platforms', *International Journal of Communication* 8: 597–617.

Liu, D., R. F. Baumeister, C. C. Yang and B. Hu. 2019. 'Digital communication media use and psychological wellbeing: A meta-analysis', *Journal of Computer-Mediated Communication* 24(5): 259–73. https://doi.org/10.1093/jcmc/zmz013.

Loos, E., G. Nimrod and M. Fernández-Ardèvol. 2018. 'Older audiences in the digital media environment: A cross-national longitudinal study', *Wave 1, Report 1.0.* Montreal, Canada: ACT project. Accessed 26 April 2020. https://spectrum.library.concordia.ca/id/eprint/983866/.

Loos, E., G. Nimrod and M. Fernández-Ardèvol. 2020. 'Older audiences in the digital media environment: A cross-national longitudinal study', *Wave 2, Report 1.0.* Montreal, Canada: ACT project. Accessed 26 April 2020. https://spectrum.library.concordia.ca/id/eprint/986444/.

Loos, E. and L. Ivan. 2022. 'Not only people are getting old, but the new media are too: Technology generations and the changes in new media use', *New Media & Society* 26(6): 3588–3613. https://doi.org/10.1177/14614448221101783.

Maderick, J. A., S. Zhang, K. Hartley and G. Marchand. 2016. 'Pre-service teachers and self-assessing digital competence', *Journal of Educational Computing Research* 54(3): 326–51. https://doi.org/10.1177/0735633115620432.

Manco-Chavez, J. A., Y. C. Uribe-Hernandez, R. Buendia-Aparcana, J. J. Vertiz-Osores, S. D. Isla Alcoser and R. A. Rengifo-Lozano. 2020. 'Integration of ICTS and digital skills in times of the pandemic COVID-19', *International Journal of Higher Education* 9(9): 11–20. https://doi.org/10.5430/ijhe.v9n9p11.

Mandache, L. A. and L. Ivan. 2022. 'Older adults and communication technologies during the lockdown in Romania'. In *Human Aspects of IT for the Aged Population. Design, interaction and technology acceptance*, edited by Q. Gao and J. Zhou, 365–80. Cham, Switzerland: Springer International Publishing. https://doi.org/10.1007/978-3-031-05581-2_27.

McLuhan, M. 1964. *Understanding Media: The extensions of man*. New York: McGraw Hill.

Midão, L., E. Pedreiro, M. S. Pinho, I. Dias, M. Almada, K. R. Garcia and E. Costa. 2020. 'Computer skills among the community-dwelling 55+ European population based on a survey of health, ageing and retirement in Europe', *International Journal of Digital Literacy and Digital Competence (IJDLDC)* 11(1): 31–45. https://doi.org/10.4018/IJDLDC.2020010102.

Ng, W. 2012. 'Can we teach digital natives digital literacy?', *Computers & Education* 59(3): 1065–78. https://doi.org/10.1016/j.compedu.2012.04.016.

Nimrod, G. 2017. 'Older audiences in the digital media environment', *Information, Communication & Society* 20(2): 233–49. https://doi.org/10.1080/1369118X.2016.1164740.

Nimrod, G. 2022. 'Technostress in a hostile world: Older internet users before and during the COVID-19 pandemic', *Aging & Mental Health* 26(3): 526–33. https://doi.org/10.1080/13607863.2020.1861213.

Nimrod, G. and L. Ivan. 2019. 'The dual roles technology plays in leisure: Insights from a study of grandmothers', *Leisure Sciences* 42: 1–18. https://doi.org/10.1080/01490400.2019.1656123.

Olsson, T., U. Samuelsson and D. Viscovi. 2019. 'Resources and repertoires: Elderly online practices', *European Journal of Communication* 34(1): 38–56. https://doi.org/10.1177/0267323118810852.

O'Sullivan, R., A. Burns, G. Leavey, I. Leroi, V. Burholt, J. Lubben and T. R. Prohaska. 2021. 'Impact of the COVID-19 pandemic on loneliness and social isolation: A multi-country study', *International Journal of Environmental Research and Public Health* 18(19): 9982. https://doi.org/10.3390/ijerph18199982.

Organisation for Economic Co-operation and Development (OECD). 2016. *Skills for a Digital World I* [report]. Accessed 15 December 2024. https://www.voced.edu.au/content/ngv:75145.

Ovanessoff, A. and M. Purdy. 2011. 'Global competition 2021: Key capabilities for emerging opportunities', *Strategy & Leadership* 39(5): 46–55. https://doi.org/10.1108/10878571111161525.

Peters, C. and K. C. Schrøder. 2018. 'Beyond the here and now of news audiences: A process-based framework for investigating news repertoires', *Journal of Communication* 68(6): 1079–1103. https://doi.org/10.1093/joc/jqy060.

Peters, C., K. C. Schrøder, J. Lehaff and J. Vulpius. 2022. 'News as they know it: Young adults' information repertoires in the digital media landscape', *Digital Journalism* 10(1): 62–86. https://doi.org/10.1080/21670811.2021.1885986.

Polanyi, M. and A. Sen. 2009. *The Tacit Dimension*. Chicago, IL: University of Chicago Press.

Postman, N. and C. Weingartner. 1969. *Teaching as a Subversive Activity*. McHenry, IL.: Delta Publishing.

Postman, N. 2000. 'The humanism of media ecology'. In *Proceedings of the Media Ecology Association*, edited by J. Sternberg and M. Lipton, vol. 1, 10–16. New York: Fordham University.

Prensky, M. 2001. 'Digital natives, digital immigrants part 2: Do they really think differently?', *On the Horizon* 9(6): 1–6. https://doi.org/10.1108/10748120110424843.

Prezioso, G., F. Ceci and S. Za. 2021. 'Is this what you want? Looking for the appropriate digital skills set'. In *Digital Transformation and Human Behavior: Lecture notes in information systems and organisation,* edited by C. Metallo, M. Ferrara, A. Lazazzara and S. Za, 69–86. Cham, Switzerland: Springer International Publishing. https://doi.org/10.1007/978-3-030-47539-0_6.

Sacks, H. 1992. *Lectures on Conversation,* vols 1–2. Edited by G. Jefferson. New York: Blackwell.

Schwarzenegger, C. 2020. 'Personal epistemologies of the media: Selective criticality, pragmatic trust and competence–confidence in navigating media repertoires in the digital age', *New Media & Society* 22(2): 361–77. https://doi.org/10.1177/1461444819856919.

Stevenson, I. 2013. 'Does technology have an impact on learning? A fuzzy set analysis of historical data on the role of digital repertoires in shaping the outcomes of classroom pedagogy', *Computers & Education* 69: 148–58. https://doi.org/10.1016/j.compedu.2013.07.010.

Taipale, S., T. A. Wilska and C. J. Gilleard (eds). 2017. 'Introduction'. *Digital Technologies and Generational Identity*, 1–9. New York: Routledge.

Taipale, S., T. Oinas and J. Karhinen. 2021. 'Heterogeneity of traditional and digital media use among older adults: A six-country comparison', *Technology in Society* 66: 148–58. https://doi.org/10.1016/j.compedu.2013.07.010.

Tarafdar, M., C. L. Cooper and J. F. Stich. 2019. 'The technostress trifecta-techno eustress, Techno distress and design: Theoretical directions and an agenda for research', *Information Systems Journal* 29(1): 6–42. https://doi.org/10.1111/isj.12169.

Tarafdar, M., C. Maier, S. Laumer and T. Weitzel. 2020. 'Explaining the link between technostress and technology addiction for social networking sites: A study of distraction as a coping behaviour', *Information Systems Journal* 30(1): 96–124. https://doi.org/10.1111/isj.12253.

Tilly, C. 2006. *Regimes and Repertoires*. Chicago, IL: University of Chicago Press.

Tomini, F., S. M. Tomini and W. Groot. 2016. 'Understanding the value of social networks in life satisfaction of elderly people: A comparative study of 16 European countries using SHARE data', *BMC Geriatrics* 16(1): 1–12. https://doi.org/10.1186/s12877-016-0362-7.

Tsekeris, C. 2019. 'Surviving and thriving in the 'Fourth Industrial Revolution': Digital skills for education and society', *Homo Virtualis* 2(1): 34–42. https://doi.org/10.12681/homvir.20192.

Zhao, L. 2021. 'The effects of mobile social media use on older migrants' social integration and life satisfaction: Use types and self-esteem perspective', *Social Science Computer Review* 41(1): 249–64. https://doi.org/10.1177/08944393211042545.

Zimbardo, P. G. 1995. 'The psychology of evil: A situationist perspective on recruiting good people to engage in anti-social acts', *Japanese Journal of Social Psychology* 11(2): 125–33. https://doi.org/10.14966/jssp.KJ00003724682.

Part II
Digital repertoires in everyday life

3
Exploring the key elements of digital repertoire: the meaning of new technology in later life

Riitta Hänninen and Sakari Taipale

Introduction

Meaningful integration of new digital technology into everyday life is a key factor in terms of perceived social and digital inclusion. What is considered meaningful, in turn, varies according to age and life stage, as well as cultural and situational factors. For example, in Finland older adults are for the most part willing to use and engage with digital technology and services, yet there is still a noticeable variation in their digital literacy and familiarity with online services (Rosenlund and Kinnunen 2018). Older adults (65+) constitute a heterogenous age group that characteristically use the internet and social media less frequently or widely than younger age cohorts do (Rasi and Taipale 2020). Furthermore, there are significantly more non-users of the internet among older adults than there are in younger age groups (Pirhonen et al. 2020; Rasi 2018). The most common reasons for using the internet among Finnish older adults are online banking, email, digital services, reading news and researching information on illnesses and health (Rasi and Taipale 2020). In terms of digital services, older adults are especially drawn to the accessibility, convenience and economic incentives associated with digital environments (Hänninen et al. 2021b; see also Hargittai et al. 2019; Seifert and Rössel 2019; Tsatsou et al. 2018).

The driving force behind our study is to explore how the concept of 'digital repertoire' can challenge the biased narrative that portrays older adults as a uniform group with limited access to digital technology (the first-level digital divide) and insufficient digital skills (the second-level digital divide). To achieve this, the study aims to identify the key elements of digital repertoires as they manifest in the everyday lives of older adults.

In our analysis we first examine the elements of digital repertoire significant in later life (1), then secondly, deriving from the concept of the third digital divide, describe what it takes for older adults to build a meaningful relationship with digital technology (2). In terms of theory, we position the digital repertoire approach in the field of ageing and digital technology studies, and in relation to the Theory of Digital Divides, Practice Theory and the Uses and Gratification Theory (UGT). The analysis is based on 20 qualitative participant-induced elicitation interviews (PIE) conducted in Central Finland in 2018. The older adults participating in the study were aged between 65 and 89. The research material was analysed following inductive thematic analysis.

Finland is a Nordic welfare country, along with Norway, Iceland and Sweden, situated in the northern hemisphere of the globe. More than 45 per cent of the 5.5 million Finnish population live alone, and approximately 33 per cent of them live in two-person families (OSF 2021a). Due to the rapid urbanisation of Finnish society, the practice of extended families with several generations living together has become scarce since the Second World War. The social distance, based on infrequent face-to-face interactions in many of the Finnish families, is further reinforced by geographical distances, which can be extensive.

Our fieldwork site in the city of Jyväskylä, for example, is located in Central Finland, 300 km from the capital city of Helsinki. The most northerly parts of Finland, situated in Lapland, are more than 1,000 km from Helsinki. In 2021 the population density in Finland was 18 inhabitants per square km, while in the European Union the same number was on average 109 (Kuntaliitto 2021). Younger family members leaving parental homes in Finland divide families relatively early into smaller units, contributing to the emergence of distributed families and making daily interactions increasingly reliant on digitally mediated family communication (Hänninen et al. 2021a). Additionally, although the standard of living in Finland is in global terms high, the same also applies to living costs. This can make it difficult for older adults relying on their pension to keep up with digitalisation.

The transition towards digital society has proceeded in Finland at a rapid pace. In 2021 the International Digital Economy and Society Index (iDESI) ranked Finland as number one in the world in terms of the performance of digital economy and society (European Commission et al. 2020). In 2010 more than half (54 per cent) of Finns used the internet several times per day. A decade later, in 2020, the corresponding figure had risen to 82 per cent of people between the ages of 16 and 89 (SF 2010; 2020). The use of digital services and social media has recently increased, especially among older adults over 75. According to the Official Statistics of Finland (OSF 2021b), 85 per cent of those aged between 65 and 74 and 57 per cent of those aged between 75 and 89 were using the internet daily or almost daily in 2021.

The ageing process in both physical and cognitive abilities can significantly affect the challenges associated with digital technology use in later life (Chang et al. 2015). Furthermore, ageing is also considered to be one of the defining elements of digital inequalities (van Dijk 2005; Urban 2017). In terms of heterogeneity, it is important to note that incentives to use, or not to use, digital technology also vary to a great extent, depending on the everyday life needs and practices of an individual older adult (Hänninen et al. 2021b; Korpela et al. 2024). We therefore claim that in order to gain understanding of both the heterogeneity of older people and the digital inequalities they encounter, it is necessary to focus on the everyday digital practices of older adults – especially the meanings they attach to these practices.

Theoretical positioning of the digital repertoire approach

According to Stevenson (2013, 156), digital repertoires consist of 'those routine practices which individuals develop when using technologies'. In terms of social theories, he positions the digital repertoire approach at the level of individual digital technology users and their everyday lives. In this chapter we take a step further, arguing that in order to become established these practices should be also considered meaningful, and that they should gratify older adults' daily needs. In Figure 3.1 we clarify the theoretical contribution of the digital repertoire approach. First, an individual's digital repertoire resides in a reciprocal relationship with structural factors that lay the foundation for digital technology use. These factors, such as digital infrastructure and economic possibilities to adopt digital technology, are typically conceptualised through the digital divide theories that imply gaps between those who have access to digital technology and those who do not (e.g. van Deursen and van Dijk 2019; Norris 2001).

In addition to these so-called first-level divides, previous research outlines second-level divides, which refer to the unequal distribution of digital skills and the actual use of digital technology (e.g. Hargittai 2002;

Hargittai et al. 2019; Scheerder et al. 2017), as well as third-level divides that allude to unequal distribution of benefits gained through skilled use of digital devices and services (e.g. Calderón Gómez 2020; van Deursen and Helsper 2015; Ragnedda 2017; Scheerder et al. 2017). While access and skills are essential parts of digital repertoires, deficiencies in them can also be compensated – for instance, by means of technology co-use (Hänninen et al. 2021b) or use by a proxy person (Dolničar et al. 2018). We therefore argue that in order to be able to benefit fully from digital technology, it is necessary that older people find practices of technology use that are suitable for them and that meet their everyday life needs.

This level of social practices and technology uses is illustrated in the bottom part of Figure 3.1. In terms of contemporary social theory, social practices are considered to evolve at the nexus of three elements – materiality, competence and meanings – that come together when a person implements a practice (Røpke 2009; Shove and Pantzar 2005). As social practices can be defined as customary ways of doing things that are developed and sustained through repetitive actions, they can be easily perceived as mundane and thus overlooked. With regards to older adults, the need for repetition, perseverance and self-paced learning are often underlined as a key to establishing new practices of digital technology use (e.g. Jin et al. 2019; Taipale 2019).

The three elements of social practices reflect the same phenomena as the digital divides, yet on the level of everyday life. The material element refers to the physical aspects of practices. It covers both the human body and non-human artefacts, such as mobile phones, laptops and internet connection that form the material preconditions for performing a practice (cf. first-level divide). Competence involves both the skills and knowledge required to undertake a practice (cf. second-level divide). The concept of meaning here is applied to underline the anticipated relevance of a practice. When initiating a new digital practice, individuals often consider its relevance in terms of the material preconditions and their own competencies. However, when an older adult decides to try a new online service, the suitability of that service is often determined by socio-economic, cultural and contextual factors that shape their ability to transform digital technology use into meaningful benefits in their lives. Known as the third-level digital divide, this emphasises the disparities in the *outcomes* or *benefits* people derive from using digital technologies, even when they have access to them and possess the skills to use them effectively.

Alongside the theories of social practice, the uses and gratification Theory (UGT) is widely used to understand people's engagement with new media and digital technology. The theory was first formulated in

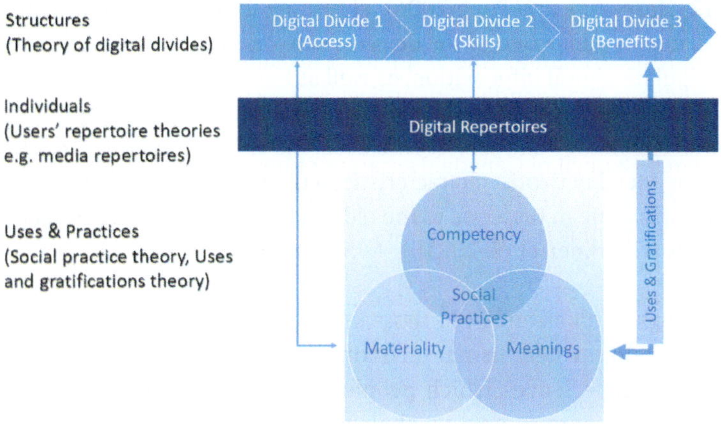

Figure 3.1 Theoretical positioning of digital repertoires. Image by the authors.

the 1940s (Katz et al. 1973) in connection with mass media audience studies. More recently it has been applied to understanding the goal-oriented adoption of digital technologies (e.g. Quan-Haase and Young 2010; Raacke and Bonds-Raacke 2008). The theory states that people adopt and use certain technologies to gratify, or meet, their individual needs, although it has been criticised for the assumptions it makes about people's ability to know their needs and to act on them effectively.

As illustrated in Figure 3.1, we propose that the need for gratification feeds into the digital repertoire approach in a particular manner. When social practices related to digital technology use are considered meaningful, people can perform particular tasks to meet their daily needs and gain benefits. However, if competencies and knowledge are insufficient, the gratification of needs may take place through 'workaround' practices. Digital services can be superseded, for example, by making a phone call to the doctor's surgery instead of using digital platforms or by asking a close relative to book an online appointment.

The scope of digital repertoires

Based on our recent study on the use of digital services (Hänninen et al. 2021b), we argue that older adults use digital technology according to their digital repertoires, which include various topics of interest, skill sets and availability of devices and applications. The scope of the concept echoes the definition of Donner (2015, 109; see also Hasebrink & Domeyer 2012; Hasebrink and Popp 2006), according to which 'a person's digital repertoire is comprised of the devices, networks and services she uses to manipulate digital information, as well as her skills to do so'. The notion of digital repertoire thus includes the devices and applications on which digital technology and services are used, the cultural and social contexts, such as social networks characteristic to the user, and the content – that is, what is done in response to the older adult's everyday needs via these devices (Hänninen et al. 2021b). As Stevenson observes:

> Digital repertoires are those routine practices which individuals develop when using technologies. They relate to the tasks and experiences which people have met in their past, but also enable them to apply technical fluency to new situations which they interpret in terms of their existing instrumented activities (2013, 156).

Although everybody using digital technology has a *de facto* digital repertoire characterised by the individual and shared (Hänninen et al. 2021b) or social aspects of digital technology use, it is important to note that ageing can have a profound effect upon the formation and transformation of one's personal digital repertoire. Firstly, older age cohorts have been found to exhibit less interest in, and use of, digital technology than younger cohorts (Jacobetty and Fernández-Ardèvol 2017; Sawchuk and Lafontaine 2020; Shultz et al. 2015), which suggests that their digital repertoires may be less flexible compared with younger users of digital technology. Secondly, the connectivity and frequency of digital technology use remain at a relatively low level among older adults (Choi and DiNitto 2013; Friemel 2016; Hänninen et al. 2021b; Nimrod 2017). Lastly, the concept of digital repertoire does not entail heterogeneity only in terms of older adults as a cohort or an age group. Rather, it underlines the diversity of digital repertoires on a personal level. The devices, applications and services used by older adults vary based not only on their age but also on their skills, interests and needs, as well as the online and offline social networks associated with their use of digital technology. Physical changes with ageing can also impact on the personal digital repertoire. They serve to reshape the needs of everyday life and thus influence the ways in which older adults perceive digital technology, both in terms of its meaning and their personal relationship with it (Charness and Boot 2009; Fernández-Ardèvol 2011; Gallistl and Nimrod 2020; Hänninen et al. 2021a; Hänninen et al. 2021b).

Touching on the concept of polymedia, 'digital repertoire' underlines what Madianou and Miller have described as the

> emerging environment of communicative opportunities that functions as an 'integrated structure' within which each individual medium is defined in relational terms in the context of all other media (Madianou and Miller 2013, 170).

Following the lines of the media use theory, digital repertoires involve relatively stable patterns of digital technology use focusing especially on the user (Hasebrink and Popp 2006) – instead of, for example, infrastructure or skill requirements that are addressed by the first- and second-level digital divide. The holistic scope of the concept emphasises the meaning of digital technology in the user's everyday life and highlights the individual lifestyle of a person in specific social and cultural contexts (Nimrod 2017; Schwarzenegger 2020).

Drawing from our qualitative interview data, we argue that the personal digital repertoire of a given person forms a complex blend of individual and social elements. The elements we examine in this study are deeply intertwined and can manifest in various, sometimes even contradictory, combinations. Digital repertoires live and evolve in the everyday life practices of older adults, including the social aspects of daily life. Therefore, especially in terms of co-use, supported use, proxy-use and non-use of digital technology (Hänninen et al. 2021a), digital repertoires can also be, or can become, influenced or even shared by other persons' digital repertoires. In this light, we argue that warm experts (Bakardjieva 2005; Hänninen et al. 2021a; Hänninen et al. 2021b), such as family members and friends who support older adults in their digital technology use, play a central role in the formation and transformation of the digital repertoires of older adults.

Research data and methods

The fieldwork focusing on older adults and the role of digital technology in their everyday lives produced 22 elicitation interviews, including one interview conducted with an elderly couple (Hänninen et al. 2021b). The PIE interviews were conducted in November and December 2018 in Central Finland. The research participants were all Finns aged between 57 and 89. In the present study we focus only on the interviewees aged 65 or over, which is, according to the United Nations, a common age for retirement in developed countries. The retirement age in Finland ranges from 63 to 68 years (FCP 2021), so 21 out of the total of 23 interviewees were included in the analysis. Twelve of the older adults participating in the research were female and nine were male. The interviews were also photographed to document the fieldwork and to give the researcher a concrete way of contextualising participants' technology use through visual references (Collier 1957; Harper 2002; Hänninen et al. 2021b; Kuoppamäki et al. 2022).

The interviewees were recruited from a housing association that provides communal housing services to older adults over the age of 55. All participants owned either a smartphone or feature phone and most also had other technological devices at their disposal, such as tablets or laptops (Hänninen 2021b; Kuoppamäki et al. 2022). The main themes of the interviews covered how and what kind of digital technology they used in their daily lives and the problems and benefits they had experienced with digital technologies and services (Kuoppamäki et al. 2022). The duration of each interview was approximately one hour.

PIE is an interview technique based on photo-elicitation (Collier 1957; Harper 2002), where the interviewee is asked to take photographs dealing with the topics of the research (Bignante 2010; Epstein et al. 2006; Hänninen et al. 2021b). In general terms, elicitation refers to a process in which a response, meaning or answer is evoked from the interviewee by using, for example, photographs. In other instances, as in this case digital devices and applications also serve as a starting point for the interview (Hänninen 2020). As many of the aspects of digitalisation are often embedded in the daily life of the person participating in the research, they become so habituated to them that it is difficult to remain consciously aware of them (Hine 2015). It was therefore helpful to have the digital devices and applications that the older adults used daily with them in the interviews. PIE also enabled us to be in methodological terms critical in situations where older adults described their use of digital technology in a different way to that in which they actually used it (see Hänninen 2020; Hänninen et al. 2021b; Kuoppamäki et al. 2022).

Initially we focused in our analysis on examining what it takes for older adults to adopt and use digital technology and services. We soon found out that the digital technology use varied greatly, depending not only on age and ageing, but also on what was particular to an individual older adult. We thus brought this observation to the centre of the analysis in order to explore the key elements of these newly discovered 'digital repertoires'.

In the first part of the thematic analysis we determined which elements of digital repertoires were recurring in the interview data and examined these elements more closely. We found that some of the key elements of digital repertoires held an individual quality to them while others operated on a distinctly social level of everyday life. In the second part of our analysis we outlined the ramifications of the individual and social (shared) aspects of digital repertoires in older adults' perspective, exploring some of the ways in which these key elements overlapped with one another within the everyday life practices of the older adults. In the following sections, we first identify the individual and social elements of digital repertoires, then move on to discuss digital repertoires on a more theoretical level.

Individual elements of digital repertoire

Many of the interviewees were initially acquainted with digital technologies in the 1990s, the decade when computer technology was first introduced to Finnish working life. Some of the interviewees, unemployed during the early 1990s recession, had also become acquainted with digital technology through computer courses, or so-called automatic data processing courses, that formed part of the mandatory labour-market oriented education for the then unemployed. The level of digital embeddedness ranged from older adults who had done office work with ICTs on a daily basis to workers who only had used timecards to record their working hours. Some of the interviewees had not used digital technology in their working lives at all, or had been exposed to digital tools only briefly before retirement.

All the interviewees seemed to benefit from encountering digital technologies in their working lives especially in terms of their motivation, skills and attitudes towards technology. Early exposure to digital technology provided them with a familiar starting point for using it, adopting new devices and applications and consequently broadening their digital repertoires in later life (see also Dutton et al. 2009; Hargittai and Dobransky 2017; Hasebrink and Domeyer 2012; Pajula et al. 2024). However, some interviewees, who can be described as 'late adopters', used digital technology regularly and in various purposes, suggesting that there are other crucial elements to the development of digital repertoire than one's working life.

The second individual key element of digital repertoire, similarly based on the life history and everyday life of the interviewees, is the complex category of 'personal everyday life'. This consists of various points of interests, digital chores and recurring practices or routines involving digital technology. Based on our research data, personal everyday life includes a variety of activities, for instance listening to digital radio every morning, checking the weather forecast from one's smartphone, learning French and other languages, watching YouTube videos about cats, dogs and turtles, playing games such as Mahjong, Hay Day and Angry Birds, solving crossword puzzles and blogging. Points of personal interest also entailed using Google Maps while foraging berries and mushrooms in the forest and elk hunting, bird watching, taking new photographs and digitalising older ones, searching for bargains from the online marketplaces and shops, engaging with genealogy, organising Tupperware parties on WhatsApp and, as in the case of Amanda, aged 84, attending opera concerts on YouTube instead of going to a live concert or listening to a record:

> Sometimes I listen it [opera music] on YouTube. [...] Recently I have been listening especially to Torikka [a Finnish opera singer]. Our family has always been musical and we have all sang in several choirs [...] I was hesitant at first when I was trying [to use YouTube] for the first few times, but then I realised that this is easy. I can do this.

Respectively, the use of digital services was typically based on the personal needs of the older adults. If the interviewee was dealing with illness, for example, it was likely that he or she was a frequent user of Omakanta (an online healthcare service for citizens of Finland). It was also common that if an older adult was using one digital service, he or she had also adopted other similar ones. However, there were also cases in which the interviewee used only one specific digital service. Marianne, aged 71, describes her situation:

> I use them [digital devices] mainly to pay bills. [...] There are all sorts of [digital services] I'm sure. And if I knew how to do them, I would. I am interested in Omakanta and many other things too. I wish I could learn how to use them.

Marianne's take on using digital services highlights an important aspect of digital repertoires. Although not everybody participating in the research was using all digital technology or services available in the market, many of them were using more than one digital service. This suggests firstly that they had a need for those particular technology and services, but also that using one technology can provide older adults with the necessary skills and encouraging experiences to reduce difficulties in adopting to other digital technologies. As Marianne's example further shows, digital repertoire is not something that would proceed from one skill, competence or aspect of digital technology to another in a straightforward manner. It is rather an intricate process that is based on the everyday life and needs of older adults and that can portray a variety of discrepancies and personal characteristics (see also Hänninen et al. 2021b). One's digital repertoire can also diminish or become more dependent on external support due to, for example, illness or a broken device that cannot be replaced with a new one because of the relatively high cost involved.

Although the previous list on the various points of interest is only a small glimpse at the rich variety of the interviewees' everyday lives, there are two kinds of everyday life activities that emerged from the research data. Firstly, a point of interest can be something that exists or once existed independently from the digital sphere of everyday life. This might include activities such as foraging for berries and mushrooms, learning new languages or listening to opera. In these types of cases the significance of digital technology stems from the ways in which it can help older adults to pursue the leisure activities they have been engaged with even before digital technology entered their everyday lives in a new way. Foraging berries and mushrooms is safer if one can use Google Maps to guide one's way in the forest. Learning new languages online instead of attending live classes can be convenient for older adults living in rural areas of Finland, while doing genealogy online can make it easier to pursue an interest in one's family history.

Secondly, based on the research data, digital technology can introduce older adults to new areas of interest or ways to manage their everyday lives – activities which they were not familiar with before, such as playing games like Hay Day and Angry Birds, blogging or sharing photographs on Instagram. Digital technologies could also bring novel topics of interest closer to older adults, thus encouraging them to try new things and broaden their digital repertoires. The research data also included examples of older adults who had adopted new ways of doing things with the assistance of digital technology, such as listening to music from one's smartphone while exercising or, as in the case of Laura, aged 74, to dispose of their old books on handicrafts and food recipes because 'nowadays everything can be found online anyway'.

In these instances we argue that digital technology affected older adults either by extending the idea of how it can contribute to their everyday life by making it more interesting, convenient and/or easier to manage, and by providing new digital infrastructure to support their wellbeing. Based on the research data, however, it was clear that if older adults did not need the digital technology or service, or were not interested in what the device or application had to offer, the digital technology was likely to be ignored, regardless of the devices or skill set of the interviewee.

The third individual element of digital repertoire consisted of the attitudes and values that older adults bestowed upon digital technology in their everyday life. In general terms digital technology was regarded as a positive thing that catered well to their daily needs. A positive attitude also endorsed the process, with existing skills being applied in new digital environments. Interestingly, however, one of the most common themes associated with digital technologies was the idea that digital technology should be used expressly for practical things, such as utilising public digital services, rather than for hobbies or leisure activities. Attitudes towards social media in particular were mixed. Several interviewees participating in the research clearly stated that they did not use social media platforms or have any interest in doing so, as they preferred face-to-face social interaction (see Leist 2013; Pirhonen et al. 2020). Although the use of digital technologies for leisure activities was certainly 'there' in the research, it was not necessarily considered to be the primary function of digital technology. This observation reveals an instrumental approach to digital technology, one that can precede the process in which digital technology becomes a more embedded part of everyday life (see also Hänninen et al. 2023). Other recurring themes reflecting the attitudes and values towards digital technology included issues associated with digital security (especially online banking services and online shopping, both of which deal with handling money online) and the idea that learning how to use digital technology is difficult in later life. Eva, aged 78, also revealed another very specific attitude concerning the supported use of digital technology:

> I am a single person without children living alone. [...] I have so many friends from my student years [in nursing school] that say they don't have to learn [how to use digital technology] because they have children who do it for them. I don't think that's okay. I must do it myself. Otherwise, I couldn't run my errands on my own.

Withdrawing oneself from the digital sphere of contemporary digital culture is a right that can be exercised by anybody who wishes to do so. In the context of ageing, however, such an attitude may also be associated with hesitance, anxiety and even fear of failing while using digital technology, suggesting that negative attitudes towards technology may have emotional origins. Furthermore, these emotions and attitudes can translate into negative meanings associated with the use of digital technology and services. In so doing, they reflect both the everyday life practices and the meanings attached to these practices by older adults (see also Fernández-Ardèvol 2011; Gallistl and Nimrod 2020; Hänninen et al. 2021a; 2021b).

Social elements of digital repertoire

According to our interview data, the social elements of digital repertoire include social or peer pressure, co-use and supported use of digital technology, and the desire to be connected with family and friends. All these elements highlight the influence of social networks in the formation of an individual digital repertoire. Firstly, digital technology is typically adopted as a result of social encouragement and/or pressure. In this example Johan, aged 82, describes his history of acquiring digital technology:

> It was my idea that I must get a smartphone because everybody seemed to have one. I thought I need to join so that I wouldn't be too different from the majority. It was the same thing with the computer too. First I thought I'm never going to need it, so I didn't get one, but [...] when I turned 82, it [tablet computer] was given to me as a birthday present [by his children].

Encouraging older adults to adopt digital technology could take place by recommending or buying them new devices, for example, or, as in the case of social pressure, through wanting to be part of the 'majority' mentioned by Johan. Social pressure also reflected upon the applications and digital services adopted and used by older adults. If the family or friends were using WhatsApp or Facebook in their internal communication, for example, it was common that the interviewees too became members of these groups. This served not only to broaden their digital repertoires, but also to explain the adoption of specific applications.

The second social element of digital repertoire is co-use and supported use (with warm experts) of digital technologies. Older adults co-used and learned how to use digital technology with the support of warm experts. In doing so, they absorbed parts of the digital repertoires of these warm experts into their own digital repertoires. Tasks involved in sharing digital repertoires with a warm expert typically ranged from acquiring devices, adopting new applications and maintaining the devices through system and security updates to solving daily problems associated with digital technology.

The social aspect of sharing could also take very practical forms, as in the case of Mikael, aged 66. Here a warm expert was involved in handing down second-hand devices to older adults:

> My brother [...] has always been interested in digital technology, laptops and what not. He's always buying newer, better devices and gives me the old ones. He gave me this Lumia phone because apparently, he found that it was not good enough for him [...] He has also given me this laptop.

The co-use of digital technology was common, especially in the context of digital services. These services provided the interviewees with an important tool to run their daily errands, such as booking a medical appointment or paying their bills, as Johan did:

> Although all my bills are paid online, I don't use digital services myself. My daughter pays the bills for me. I haven't had to learn how to do it. [...] It's too difficult for me so I have skipped it [...]. But now everything is running smoothly, and I save a lot of money in service fees.

The supported use of digital technology and services varied from supported use and co-use (with warm experts) to non-use of services among the older adults participating in the study. This diversity served to highlight the collaborative and social basis of sharing digital repertoires (Hänninen et al. 2021a; 2021b). Such 'sharing' could in fact be based on giving advice and doing things together (co-use). Alternatively, as in Johan's case, sharing could also take a more direct form of using digital technology on behalf of the older adult (use by a proxy).

The third social element of the digital repertoire underlines the significance of communication and connectedness. New devices and applications were adopted by the interviewees in order to keep in touch with their children and other relatives who did not live nearby or had moved abroad to pursue work or studies. In some cases, adopting new digital technology had been prompted by an unanticipated change within the family. Frans, for example, aged 72, purchased his first mobile phone because his mother became ill. He had to have the mobile phone on him at all times, so that he could be reached if necessary. Another incentive to adopt digital technology was to stay in contact with one's grandchildren, friends and other acquaintances, or to participate in a hobby group or a work group.

Both individual and social elements of digital repertoires in many ways overlapped with one another. They also occurred in various combinations, depending on the personal digital repertoire of the older adult. For example, older adults could share their leisure activities with their family and friends, portraying an individual element of personal interest and a social element of communication and connectedness. Previous working life and the supported use of digital technology can also overlap with several other elements of one's digital repertoire, including, for example, the attitudes and values associated with digital technology use.

Conclusions and discussion

In this chapter we have argued that older adults adopt and use digital technology according to their personal digital repertoires. These include diverse topics of interest, skill sets and availability of devices and applications (see also Hänninen et al. 2021b). Theoretically we positioned digital repertoires with relation to the Theory of Digital Divides, Practice Theory and the Uses and Gratification theory (UGT). In particular, we focused upon an investigation of the elements of digital repertoire among Finnish research participants that are significant in their later life. In addition, we described what it takes for older adults to build a meaningful relationship with digital technology. In the light of this analysis we identified three individual elements (working life, everyday life personal preferences, attitudes and values) and three social elements (social encouragement and/or pressure, co-use/supported use of digital technology, and communication and connectedness) of a person's digital repertoire.

In accordance with the first research question, focusing on the individual and social elements of digital repertoires in later life, we revealed that using digital technology – rather than relying on access and skills, as suggested by the first-level and second-level digital divides – reflects the significance of personal everyday life. This is particularly clear in the connection to previous working life, personal preferences, and attitudes and values associated with digital technology use.

The social aspects of digital repertoire broaden our analysis on personal digital repertoires by adding the warm experts, such as family and friends, to this equation. Social encouragement and/or pressure, as well as co-use, supported use and communication/connectedness, complement the digital repertoire of an older adult and provide assistance and new dis/incentives to use digital technology. Based on our analysis, we also found that both the individual and social elements of digital repertoires tend to overlap, which contributes to understanding of the heterogeneity of technology use in later life.

The close relationship between both individual and social elements of digital repertoire constitutes an important new direction for future research. It emphasises that digital technology use is rarely guided only by an older person's rational thinking (as suggested by UGT). Instead it is influenced by a number of social and cultural factors that add to the meaningfulness of digital practices (as suggested by social practice theory) in later life. It is, in fact, this complex combination of individual and social factors that contribute to the development of individual digital repertoires.

Following the second research question on what it takes for older adults to build a meaningful relationship with digital technology, we argue that the concept of digital repertoire acknowledges the notion of need in the context of digital technology use. If older adults do not find a certain device, application or network helpful, or interesting in their daily lives, it is more likely that they choose not to adopt or use it. Additionally, digital repertoire as a concept addresses the heterogeneity of digital technology use – not only in terms of ageing, but also in various other societal contexts relevant to digital technology use.

In terms of social theory, this chapter showed that the concept of digital repertoire bridges the gap between macro-level theories, such as the digital divide theory that describe structural preconditions for digital engagement, and the theories that operate at the level of individual technology practices, such as UGT. In so doing, the concept of digital repertoire can make sense of the everyday life of an individual technology user, acknowledging the important role played by a user's immediate social networks in the context of digital repertoires. Regarding older adults, we consider that this conceptual approach is of particular value. In later life people's daily needs, social networks and technology interests, to say nothing about physical and technical capabilities, vary tremendously. In many previous theoretical approaches this heterogeneity is easily overlooked by focusing on the dividing, macro-level categorisations, such as age divides or age cohorts, or by zooming into the particular uses of specific technology.

The current study also revealed new questions that require further investigation in the future. First, our interview data does not fully explain why some older adults manage to expand their digital repertoires from, for example, one digital service to another, but others do not. We believe that by studying in more detail the interconnections between the various elements of digital repertoire in different countries and cultural contexts may clarify this further. In addition, the interviews also indicate that the role of warm experts in older adults' personal digital repertoires may be more complicated than has been described above. Although the help received from warm experts seems generally to be appreciated, it may also have adverse effects on older adults' learning and self-determination. Our findings thus call for a more critical approach to the study of the relationship between older adults and warm experts.

Acknowledgements

This work was supported by the Research Council of Finland's Centre of Excellence in Research on Ageing and Care (grant numbers 312367 and 336671), Strategic Research Council of Finland (grant numbers 327145, 352501, 352505 and 327149) and it was conducted in partnership with the Aging in Data (AiD) project (SSHRC Partnership Grant).

References

Bakardjieva, M. 2005. *Internet Society: The internet in everyday life*. London: Sage Publications Ltd. https://doi.org/10.4135/9781446215616.

Bignante, E. 2010. 'The use of photo-elicitation in field research. Exploring Maasai representations and use of natural resources', *EchoGéo 11*. https://journals.openedition.org/echogeo/11622.

Calderón Gómez, D. 2020. 'The third digital divide and Bourdieu: Bidirectional conversion of economic, cultural, and social capital to (and from) digital capital among young people in Madrid', *New Media & Society* 23(9): 2534–53. https://doi.org/10.1177/1461444820933252.

Chang, J., C. McAllister and R. McCaslin. 2015. 'Correlates of, and barriers to, internet use among older adults', *Journal of Gerontological Social Work* 58(1): 66–85. https://doi.org/10.1080/01634372.2014.913754 .org/10.1080/01634372.2014.913754.

Charness, N. and W. R. Boot. 2009. 'Ageing and information technology use: Potential and barriers', *Current Directions in Psychological Science* 18(5): 253–8. https://doi.org/10.1111/j.1467-8721.2009.01647.x.

Choi, N. G. and D. N. DiNitto. 2013. 'The digital divide among low-income homebound older adults: Internet use patterns, eHealth literacy and attitudes toward computer/internet use', *Journal of Medical Internet Research* 15(5): e93. https://doi.org/10.2196/jmir.2645.

Collier, J. 1957. 'Photography in anthropology: A report on two experiments', *American Anthropologist* 59(5): 843–59. https://doi.org/10.1525/aa.1957.59.5.02a00100.

van Deursen, A. and E. J. Helsper. 2015. 'The third-level digital divide: Who benefits most from being online?', *Communication and Information Technologies Annual (Studies in Media and Communications*, vol. 10), 29–52. Leeds: Emerald Group Publishing Limited. https://doi.org/10.1108/S2050-206020150000010002.

van Deursen, A. and J. van Dijk. 2019. 'The first-level digital divide shifts from inequalities in physical access to inequalities in material access', *New Media & Society* 21(2): 354–75. https://doi.org/10.1177/1461444818797082.

van Dijk, J. A. 2005. *The Deepening Divide: Inequality in the information society*. Thousand Oaks, CA: Sage Publications.

Dolničar, V., D. Grošelj, M. F. Hrast, V. Vehovar and A. Petrovčič. 2018. 'The role of social support networks in proxy internet use from the intergenerational solidarity perspective', *Telematics and Informatics* 35(2): 305–17. https://doi.org/10.1016/j.tele.2017.12.005.

Donner, J. 2015. *After Access: Inclusion, development and a more mobile internet*. Cambridge, MA: MIT Press.

Dutton, W., E. Helsper and M. Gerber. 2009. *The Internet in Britain 2009*. Oxford: University of Oxford.

Epstein, I., B. Stevens, P. McKeever and S. Baruchel. 2006. 'Photo elicitation interview (PEI): Using photos to elicit children's perspectives', *International Journal of Qualitative Methods* 5(3): 1–11. https://doi.org/10.1177/160940690600500301.

European Commission: Directorate-General for Communications Networks, Content and Technology. Foley, P., D. Sutton, R. Potter, S. Patel et al. 2020. *International Digital Economy and Society Index 2020 – Final report*. European Commission Publications Office, 2020. Accessed 13 December 2024. https://data.europa.eu/doi/10.2759/757411.

Fernández-Ardèvol, M. 2011. 'Mobile phones in a retirement home: Strategic tools for mediated communication'. In *Digital Information and Communication Technology and its Applications* (Communications in Computer and Information Science), edited by H. Cherifi, J. M. Zain and E. El-Qawasmeh, Vol. 166, 395–406. Cham, Switzerland: Springer. https://doi.org/10.1007/978-3-642-21984-9_34.

Finnish Centre for Pensions (FCP). 2021. *Retirement Ages*. https://www.etk.fi/en/work-and-pensions-abroad/international-comparisons/retirement-ages/.

Friemel, T. N. 2016. 'The digital divide has grown old: Determinants of a digital divide among seniors', *New Media & Society* 18(2): 313–31. https://doi.org/10.1177/1461444814538648.

Gallistl, V. and G. Nimrod. 2020. 'Media-based leisure and wellbeing: A study of older internet users', *Leisure Studies* 39(2): 251–65. https://doi.org/10.1080/02614367.2019.1694568.

Hänninen, R. 2020. 'Participant-induced elicitation in digital environments'. In *Ethnography with a Twist. Methodological and ethical challenges and solutions in contemporary research*, edited by T. Lähdesmäki, E. Koskinen-Koivisto, V. Čeginskas and A.-K. Koistinen, 55–67. London: Routledge.

Hänninen, R., S. Taipale and A. Korhonen. 2021a. 'Refamilisation in the broadband society – The effects of ICTs on family solidarity in Finland', *Journal of Family Studies* 27(2): 161–77. https://doi.org/10.1080/13229400.2018.1515101.

Hänninen, R., L. Pajula, V. Korpela and S. Taipale. 2021b. 'Individual and shared digital repertoires: Older adults managing digital services', *Information, Communication and Society* 23(3): 568–58. https://doi.org/10.1080/1369118X.2021.1954976.

Hänninen, R., V. Korpela and L. Pajula. 2023a. 'The paradoxes and pragmatics of digital leisure in later life', *Leisure Studies*, 1–14. https://doi.org/10.1080/02614367.2023.2271194.

Hargittai, E. 2002. 'Second-level digital divide: Differences in people's online skills', *First Monday* 7(4). https://doi.org/10.5210/fm.v7i4.942.

Hargittai, E. and K. Dobransky. 2017. 'Old dogs, new clicks: Digital inequality in skills and uses among older adults', *Canadian Journal of Communication* 42(2): 195–212. https://doi.org/10.22230/cjc.2017v42n2a3176.

Hargittai, E., A. M. Piper and M. R. Morris. 2019. 'From internet access to internet skills: Digital inequality among older adults', *Universal Access in the Information Society* 18: 881–90. https://doi.org/10.1007/s10209-018-0617-5.

Harper, D. 2002. 'Talking about pictures: A case for photo elicitation', *Visual Studies* 17(1): 13–26. https://doi.org/10.1080/14725860220137345.

Hasebrink, U. and H. Domeyer. 2012. 'Media repertoires as patterns of behaviour and as meaningful practices. A multimethod approach to media use in converging media environments', *Participations: Journal of Audience & Reception Studies* 9(2): 757–79.

Hasebrink, U. and J. Popp. 2006. 'Media repertoires as a result of selective media use. A conceptual approach to the analysis of patterns of exposure', *Communications* 31(3): 369–87. https://doi.org/10.1515/COMMUN.2006.023.

Hine, C. 2015. *Ethnography for the Internet: Embedded, embodied and everyday*. London: Bloomsbury Publishing.
Jacobetty, P. and M. Fernández-Ardèvol. 2017. *Older Audiences and Digital Media: Spain, 2016. Preliminary Results* [report]. http://www.communicationchange.net/en/older-audiences-2016/.
Jin, B., J. Kim and L. M. Baumgartner. 2019. 'Informal learning of older adults in using mobile devices: A review of the literature', *Adult Education Quarterly* 69(2): 120–41. https://doi.org/10.1177/0741713619834726.
Katz, E., J. G. Blumler and M. Gurevitch. 1973. 'Uses and gratifications research', *Public Opinion Quarterly* 37(4): 509–23. https://doi.org/10.1086/268109.
Korpela, V., L. Pajula and R. Hänninen. 2024. 'Investigating the multifaceted role of warm experts in enhancing and hindering older adults' digital skills in Finland', *International Journal of Lifelong Education* 43(5): 509–22. https://doi.org/10.1080/02601370.2024.2353176.
Kuntaliitto. [The Association of Finnish Cities and Municipalities]. 2021. *Kuntien pinta-alat ja asukastiheydet* [Surface areas and population densities of municipalities]. Accessed 26 February 2024. https://www.kuntaliitto.fi/tilastot-ja-julkaisut/kaupunkien-ja-kuntien-lukumaarat-ja-vaestotiedot/kuntien-pinta-alat-ja-asukastiheydet.
Kuoppamäki, S., R. Hänninen and S. Taipale. 2022. 'Enhancing digital inclusion through social support among older adults'. In *Digital Inclusion: Enhancing vulnerable people's social inclusion and welfare?*, edited by P. Tastsou, 211–30). Cham, Switzerland: Springer International Publishing.
Leist, A. K. 2013. 'Social media use of older adults: A mini-review', *Gerontology* 59(4): 378–84. https://doi.org/10.1159/000346818.
Madianou, M. and D. Miller. 2013. 'Polymedia: Towards a new theory of digital media in interpersonal communication', *International Journal of Cultural Studies* 16(2): 169–87. https://doi.org/10.1177/1367877912452486.
Nimrod, G. 2017. 'Older audiences in the digital media environment', *Information, Communication & Society* 20(2): 233–49. https://doi.org/10.1080/1369118X.2016.1164740.
Norris, P. 2001. *Digital Divide: Civic engagement, information poverty, and the internet worldwide.* Cambridge: Cambridge University Press.
Official Statistics of Finland (OSF). 2021a. *Dwellings and housing conditions* [e-publication]. Statistics Finland. Accessed 18 February 2022. https://www.stat.fi/til/asas/index_en.html.
Official Statistics of Finland (OSF). 2021b. *Use of information and communications technology by individuals* [e-publication]. Statistics Finland. Accessed 18 February 2022. http://www.stat.fi/til/sutivi/index_en.html.
Pajula, L., R. Hänninen, V. Korpela and S. Taipale. 2024. 'Exploring stakeholder perspectives on user involvement in designing digital public services for older adults', *First Monday* 29(9). https://doi.org/10.5210/fm.v29i9.13330.
Pirhonen, J., L. Lolich, K. Tuominen, O. Jolanki and V. Timonen. 2020. '"These devices have not been made for older people's needs" – Older adults' perceptions of digital technologies in Finland and Ireland', *Technology in Society* 62, Article 101287. https://doi.org/10.1016/j.techsoc.2020.101287.
Quan-Haase, A. and A. L. Young. 2010. 'Uses and gratifications of social media: A comparison of Facebook and instant messaging', *Bulletin of Science, Technology & Society* 30(5): 350–61. https://doi.org/10.1177/0270467610380009.
Raacke, J. and J. Bonds-Raacke. 2008. 'Myspace and Facebook: Applying the uses and gratifications theory to exploring friend-networking sites', *CyberPsychology & Behavior* 11(2): 169–74. https://doi.org/10.1089/cpb.2007.0056.
Ragnedda, M. 2017. *The Third Digital Divide: A Weberian approach to digital inequalities.* London: Routledge.
Rasi, P. 2018. 'Internet nonusers'. In *The SAGE Encyclopedia of the Internet*, edited by B. Warf, vol. 2, 2532–9. London: SAGE Publications. https://doi.org/10.4135/9781473960367.
Rasi, P. and S. Taipale. 2020. 'Tuki, ohjaus ja koulutus – ikääntyneet digitalisoituvassa mediayhteiskunnassa' [Support, guidance and training – older people in a digitalising media society], *Gerontologia* 34(4): 328–32. https://journal.fi/gerontologia/article/view/99601.
Røpke, I. 2009. 'Theories of practice: New inspiration for ecological economic studies on consumption', *Ecological Economics* 68(10): 2490–7. https://doi.org/10.1016/j.ecolecon.2009.05.015.

Rosenlund, M. and U.-M. Kinnunen. 2018. 'Ikäihmisten kokemukset terveydenhuollon sähköisten palvelujen käytöstä ja kokemusten hyödyntäminen palvelujen kehittämisessä – kuvaileva kirjallisuuskatsaus' ['Older adults' experiences on digital health care services in the context of service development: A descriptive literacy review'], *Finnish Journal of eHealth and eWelfare* 10(2–3): 264–84. https://erepo.uef.fi/handle/123456789/6764.

Sawchuk, K. and C. Lafontaine. 2020. '"Just because you're older…": Seniors respond to telecom sales practices in Canada'. In *Human Aspects of IT for the Aged Population. Technologies, design and user experience*, edited by Q, Gao and J. Zhou, 231–44. HCII 2020. Lecture Notes in Computer Science, 12207. Cham, Switzerland: Springer International Publishing. https://doi.org/10.1007/978-3-030-50252-2_18.

Scheerder, A., A. van Deursen and J. van Dijk. 2017. 'Determinants of internet skills, uses and outcomes. A systematic review of the second- and third-level digital divide', *Telematics and Informatics* 34(8): 1607–24. https://doi.org/10.1016/j.tele.2017.07.007.

Schwarzenegger, C. 2020. 'Personal epistemologies of the media: Selective criticality, pragmatic trust, and competence–confidence in navigating media repertoires in the digital age', *New Media & Society* 22(2): 361–77. https://doi.org/10.1177/1461444819856919.

Seifert, A. and J. Rössel. 2019. 'Digital participation'. In *Encyclopedia of Gerontology and Population Aging*, edited by D. Gu and M. Dupre, 1446–50. Cham, Switzerland: Springer. https://doi.org/10.1007/978-3-030-22009-9_1017.

Shove, E. and M. Pantzar. 2005. 'Consumers, producers and practices: Understanding the invention and reinvention of Nordic walking', *Journal of Consumer Culture* 5(1): 43–64. https://doi.org/10.1177/1469540505049846.

Shultz, R., H.-W. Wahl, J. T. Matthews, A. De Vito Dabbs, S. R. Beach and S. J. Czaja. 2015. 'Advancing the aging and technology agenda in gerontology', *The Gerontologist* 55(5): 724–34. https://doi.org/10.1093/geront/gnu071.

Statistics Finland (SF). 2010. *'Väestön tieto- ja viestintätekniikan käyttö, Liitetaulukko 4. Internetin käyttö, käytön useus sekä määrä iän, toiminnan ja asuinpaikan kaupunkimaisuuden mukaan 2010'*, Statistics Finland. Accessed 26 February 2024. https://www.stat.fi/til/sutivi/2010/sutivi_2010_2010-10-26_tau_004_fi.html.

Statistics Finland (SF). 2020. *'Väestön tieto- ja viestintätekniikan käyttö'*. Statistics Finland. Accessed 26 February 2024. https://www.stat.fi/til/sutivi/2020/sutivi_2020_2020-11-10_tie_001_fi.html.

Stevenson, I. 2013. 'Does technology have an impact on learning? A "Fuzzy Set Analysis" of historical data on the role of digital repertoires in shaping the outcomes of classroom pedagogy', *Computers & Education* 69: 148–58. https://doi.org/10.1016/j.compedu.2013.07.010.

Taipale, S. 2019. *Intergenerational Connections in Digital Families*. Cham, Switzerland: Springer International Publishing.

Tsatsou, P., S. Kurnia and J, Choudrie. 2018. 'Introduction'. In *Social Inclusion and Usability of ICT-enabled Services*, edited by J. Choudrie, S. Kurnia and P. Tsatsou, 11–18. New York, NY, Routledge.

Urban, M. 2017. 'Embodying digital ageing: Ageing with digital health technologies and the significance of inequalities'. In *Precarity Within the Digital Age*, edited by B. Heidkamp and D. Kergel, 163–78. Cham, Switzerland: Springer International Publishing. https://doi.org/10.1007/978-3-658-17678-5_11.

4

Lives on hold? Ageing migrants' affective digital repertoires during lockdowns in Victoria, Australia

Earvin Charles Cabalquinto

Introduction

> This technology, it helped a lot. So it was just the right timing as well because the time that people are in isolation, you can still connect. [...] Even though we don't have concerts, and we cannot go out dancing, and we cannot go out and do all these activities, we can do it with this technology even if we are not physically connected. So it's fantastic. So that's the good thing with the technology. We can use it in a very good way.

One sunny afternoon I had the opportunity to interview Mercy, a 62-year-old Filipina, via Zoom. The quote above is from her, a joyful response that emphasises the benefits of digital media use in forging and sustaining connections at a time when meeting up physically was impossible. During the interview in 2020 she was living in her own home, with two international students renting rooms. Mercy had moved to Australia in 1988 and married an Australian. The marriage lasted for 17 years, ending in 2004. She and her husband did not have any children.

Currently Mercy works in the aged care and disability sector. During lockdown in 2020 she mainly used her work laptop to connect with her clients. However, in some cases she visited them physically to deliver in-person care and support. In terms of her technological profile, she owned and used a smartphone, an iPad and a laptop. She also used Facebook, Facebook Messenger, Zoom and YouTube. These online platforms enabled her to connect with her peers in Victoria, as well as her 'left-behind' family members in the Philippines. On top of these online channels she had a smart television in her home, allowing her to watch a multitude of YouTube videos. YouTube allowed her to access Hallmark movies, whose diverse themes and genre fed her appetite for entertaining content. In 2021 she started a Netflix subscription, giving her more options to watch series and movies. Indeed, an array of digital technologies and online channels gave Mercy the tools and content to stay connected while navigating the lockdown. She had the material access, resources to cover connectivity costs and the requisite technological skills to experience positive outcomes in using digital technologies during the pandemic. However, this was not case for some ageing migrants. This study will illustrate this reality.

This chapter investigates the ways in which 15 older adults from Culturally and Linguistically Diverse (CALD) backgrounds[1] used communication technologies and online networked platforms to navigate a series of lockdowns in Victoria, Australia. Victoria had six lockdowns between 2020 and 2021 during the Covid-19 pandemic. In 2020 people were allowed to go out of their homes for four reasons, including buying basic goods, exercising, delivering in-person care and to pursue work and study where it was impossible to do so from home. In 2021 leaving home to receive a vaccination was added to the rules. Importantly, Australia is home to millions of ageing CALD people. Based on the most recent report by the Australian Bureau of Statistics (2021), there were 7.6 million migrants living in Australia. Based on this figure 2.9 per cent were male and 3.1 per cent were female aged 65 and over (Australian Bureau of Statistics 2021).

The study deploys the digital repertoire perspective (Hänninen et al. 2021a) to analyse data collected through the remote interviewing of 15 participants – that is, via a phone call and/or a Zoom meeting. Ten of these participants were involved in follow-up interview sessions. Through a digital repertoire perspective, the study focuses on how physical, contextual, situational and technological factors shape the heterogenous digital practices of older adults (Hänninen et al. 2021a). However, this study interrogates and advances this perspective by illuminating how socio-cultural and technological factors produce the affective dimension of digital media use among ageing migrants during a series of lockdowns in Victoria, Australia. This point echoes the need to unpack how cultural and contextual domains – performing cultural identities, maintaining connectedness and belonging – influence the digital behaviours of older adults (Fernández-Ardèvol 2020), particularly of ageing migrants (Baldassar and Wilding 2019; Garattini and Prendergast 2017; Lamb 2020; Sokolovsky 2020) during a global pandemic.

I consider affective experiences to be those mediated by digital technologies, technological affordances and usage (Lasén 2004). I reflect on affect as a form of energy or vibe that is mediated through objects and technologies (Ash 2015) as well as cultural systems and values (Holmes and Wilding 2019). In a digital world, affect is produced through the circulation of various information through digital media use in both a local (Lünenborg and Maier 2018) and transnational context (Leurs 2014; Wise and Velayutham 2017). Importantly, informed by cultural and political differences (Ahmed 2014), it is integral in the mediation, production, embodiment and negotiation of various emotions (Lünenborg and Maier 2018; Witteborn 2014).

In unpacking the affective dimensions of ageing migrants' digital repertories, the study has been guided by the following inquiries: What is the role of digital communication technologies and online platforms in shaping the ways in which ageing migrants maintain a sense of co-presence despite forced physical separation because of lockdowns in 2020 and 2021? What influences engender or undermine digital repertoires of ageing migrants? How do a range of socio-cultural systems and technological factors impact the affective dimensions of mediated communication? What lessons can be learned from investigating the quality and dynamics of (dis)connections during a global health crisis?

In addressing these questions, this chapter showcases the benefits and challenges of digital media use. These outcomes reflect the affective digital experiences of ageing migrants as shaped by issues of uneven digital access, technological competencies and accessibility of support networks to fix technical issues. Nonetheless, it also underlines a digital repertoire lens as a constant negotiation of connections (Hjorth et al. 2018; Sinanan 2019), especially among migrants and their distant networks (Cabalquinto 2022a; Leurs 2014; Patterson and Leurs 2020). Paying close attention to the dynamics and textures of (dis)connective practices provides a critical starting point to map and examine how certain groups, such as ageing migrants, can become at risk of being left behind in an increasingly digital society, both in and beyond pandemic times.

Affective digital repertoires in enabling imagined co-presence

In the last decade studies have underlined the vital role of rapidly evolving mobile communication technologies in facilitating dispersed individuals to enact mediated co-presence. For instance, the sending of mundane text messages, photos and multimedia contents contributes to being co-present with distant others (Ling 2008). Different typologies of enacting mediated co-presence have been proposed by mobile communication scholars, including 'connected presence' (Licoppe 2004), 'ambient co-presence' (Madianou 2016), 'visual co-presence' (Ito 2005) and 'telepresence' (Richardson and Wilken 2012). These terms highlight the everyday connective practices of dispersed individuals as shaped by the features and affordance of mobile devices and social media platforms. It is worth noting that types of mediated co-presence are shaped by technological affordances (Hutchby 2014).

More recently, the mainstreamed uptake of ubiquitous information and communication technologies has also highlighted the diversity of ways a sense of connected co-presence is experienced in a transnational domain (Madianou and Miller 2012; Nedelcu and Wyss 2016). Specifically, studies have shown how transnational connections facilitated by broadband-based technologies and channels contribute to ways of staying connected and therefore embodying co-presence at a distance (Baldassar 2008; Cabalquinto 2022a; Madianou and Miller 2012; Nedelcu and Wyss 2016). Scholars have argued that this collapse of distance through digital media use sustains familial connections and a sense of home (Boccagni 2022; Cabalquinto 2022a). Indeed, digital technologies operate as critical tools to develop intimate relations and social capital (Leurs 2014; Nedelcu 2012).

Studies have underscored the role of digital technologies in facilitating a sense of co-presence among older adults and both their local (Hänninen et al. 2020; Hänninen et al. 2021a) and transnational networks (Baldassar 2008; Baldassar and Wilding 2019; Baldassar et al. 2020). Several studies have highlighted how 'frequent phone calls' and video calls facilitate the maintenance of familial and caring mediated co-presence between older people and their family members overseas (Ahlin 2020). Constant phatic communication – i.e. communication in which interpersonal contact is more important than the content of conversations – is key to generating a sense of co-presence as well as reassuring dispersed family members of their welfare (Ahlin 2020). During the pandemic older adults used digital technologies to sustain ties while navigating the lockdown (Moore and Hancock 2020). This has also been the case for older migrants who heavily rely on digital connections to maintain ties among their networks, both local and overseas. For instance, older Chinese migrants in Australia use WeChat and WhatsApp to connect to local and transnational networks (S. Wang 2024; W. Y. Wang 2024), thus paving the way to combat isolation.

In some cases digital technology use enables emotional virtual co-presence among immigrant grandchildren and grandparents in Poland (Popyk and Pustulka 2021).

Complementing these studies, this chapter underscores the digital practices of older adults in a digitised society (Fernández-Ardèvol 2020; Hänninen et al. 2020; Hänninen et al. 2021b) in a migration and ageing context (Baldassar and Wilding 2019; Baldassar et al. 2020). It also explores how cultural systems (Garattini and Prendergast 2017; Lamb 2020; Sokolovsky 2020) impact the digital behaviours of ageing migrants (Baldassar and Wilding 2019) while navigating stay-at-home orders during the pandemic.

The study deploys the digital repertoire perspective (Hänninen et al. 2021a), determining the heterogenous practices of older adults in using communication technologies as shaped by physical, contextual, situational and environmental factors (Hänninen et al. 2021a). For example, in examining the case of Finnish households, Taipale (2019) has highlighted how familial values and expectations produce the quality of digital exchanges among intergenerational household members. Furthermore, familial structures and technological capacities mobilise the exchange of technological knowledge and provision of technical support in an intergenerational familial setting (Hänninen et al. 2021b).

In addition, the digital behaviours of older migrants are moulded by their digital environment, personal and social relations and support networks. A case in point is how older Chinese migrants in Australia connect with their family members and peers, both in the country and overseas, through the networked affordance of WeChat and WhatsApp (Wang, W. Y. 2024). Importantly, the obligation to enact care among family members manifests in the ways in which older migrants connect with their grandchildren, children and other relatives on WhatsApp (Popyk and Pustulka 2021). Such practices reflect a form of digital kinning, enabling the maintenance of familial ties and relationships via digital technology use. This chapter builds on studies that unpack the emotive aspect of digital repertoires during the pandemic (Popyk and Pustulka 2021; Wang, S. 2024).

In identifying how socio-cultural and technological factors impact digital media use, the study also illuminates its affective dimensions. In the first instance affect is understood as an energy or vibe that often flows through objects and technologies (Ash 2015). More recently it has been extended to include assessment of the affects produced through digital media use among migrants (Leurs 2014; Wise and Velayutham 2017). Building on the work of Sara Ahmed (2014) on affective economies, Leurs (2014) has uncovered the affective experiences of migrants in using digital technologies to traverse physical containment in a physical space and separated from their family members. For him, affective experiences emanate from migrants' interactions with the digital screen and information; they pave the way for them to feel a sense of comfort, belonging and ontological security in uncertain conditions (Leurs 2014).

Notably, the proposition of Leurs (2014) on affective digital media use can be linked to an understanding that mobile communication technologies are affective tools (Lasén 2004) in facilitating mediated co-presence (Vincent and Fortunati 2009). The use of mobile technologies serves to generate the flow of energies that trigger the performance, embodiment and negotiation of a diverse range of emotions (Sinanan 2019; Tettegah 2016; Witteborn 2014). Importantly, technologies and infrastructures do not solely dominate the production of affect. Scholars have argued that social relations and cultural expectations mediate the affective experiences of connectivity (Holmes and Wilding 2019; Madianou and Mille 2012; Wise and Velayutham 2017).

This chapter interrogates how ageing migrants' affective experiences in using digital technologies have been shaped by the intertwining of socio-cultural and technological factors (Alinejad 2019; Cabalquinto 2022a; Leurs 2014). Previous studies on problematising co-presence at a distance have underlined how uneven technological access and digital produce (dis)connections among migrants (Cabalquinto 2022a; Leurs 2014; Madianou and Miller 2012; Parreñas 2005; Wilding 2006). Furthermore, socio-cultural expectations, power relations and uneven living conditions mediated differential digital media use and non-use (Alinejad 2019; Baldassar and Merla 2014; Cabalquinto 2022a; Lim 2016; Madianou and Miller 2012; Parreñasa 2005; Twigt 2022). To manage the frustrations, unsettling experiences and disconnections, migrants and their distant networks constantly negotiate staying connected, such as enduring unstable connections or suppressing emotions (Cabalquinto 2022a; Leurs 2014; Twigt 2022) and carefully curating and visibilising online contents (Alinejad 2019; Cabalquinto 2022a; Leurs 2014; Sampaio 2020; Twigt 2022).

During the pandemic, older migrants experienced frustrations when accessing information through a small screen device (Cabalquinto 2022c) and foreign health information resources (Wang, W. Y. 2024). In addition, virtual proximity among older migrants and their distant networks does not necessarily translate to psychological connectedness, especially when the older migrant has no deep affinity towards a particular relative (Popyk and Pustulka 2021). These digital behaviours demonstrate the diversity of accessing and using digital technologies as shaped by differing access, competencies and structural constraints.

Echoing the point of Sara Ahmed (2014), I highlight the politics of affective experiences of digital media use as shaped by disproportionate social and technological factors. Situating this inquiry within the growing body of scholarship on the way in which migrants (Cabalquinto 2022a; Cuban 2018; Horst 2011; Peile 2018) use digital technologies to manage a crisis provides a vantage point to unpack critically the affective digital repertoires of ageing migrants. Studies have shown how migrants send money (Cuban 2018; Peile 2018) or personal messages and call frequently (Cabalquinto 2022a; Horst 2011) to express support during challenging times. Nevertheless, this chapter focuses on the affective digital repertoires of ageing migrants during the 2020 and 2021 lockdown in Victoria, Australia. In doing so it complements a growing research on the heterogeneous digital practices of older migrants during a global health crisis (Popyk and Pustulka 2021; Wang, S. 2024; Wang, W. Y. 2024).

Methods of investigation

The findings were drawn from a project designed to examine how older people from Culturally and Linguistically Diverse (CALD) backgrounds accessed and used digital communication technologies during the Covid pandemic. More specifically, the project involved 15 elderly CALD people in Victoria, Australia, recruited via contacting different migrant organisations advocating for the social welfare of ageing migrants in Victoria. The recruitment stage took place in August 2020, at a time when Victoria was put into a stringent lockdown. As a result, the circulation of a call-for-participants was made via email among relevant migrant organisations, Facebook pages and Twitter.[2] Several research participants expressed their interest via an email and a phone call. The study obtained ethics approval from the Deakin University Human Research Ethics Committee (DUHREC) with number HAE-20-106.

The research had diverse participants, ranging in age from 60 up to 96. Nine of the 15 research participants were women and six were men. The majority had completed a university degree; only two had not. Twelve had already retired and three were still working. Ten were living with their family members and five were living alone. In terms of geographical background, 13 were originally born within Asia and the Pacific region, one was from Europe and one came from the Middle East.

The study employed qualitative interview techniques.[3] Due to the stay-at-home orders, interviews, each with a duration of 45 to 120 minutes, were conducted remotely, from September to October 2020. Ten of the 15 participants were engaged in a follow-up interview in August 2021, each session lasting 30 to 60 minutes. Building on a thematic approach towards analysing migration pathways and histories (Baldassar and Wilding 2019; Baldassar et al. 2020) in a digital and transnational context (Baldassar and Wilding 2019; Cabalquinto 2020; Madianou and Miller 2012), the interview questions focused on the participants' history of migration to Australia, digital media and online platform access and use, current living arrangements, benefits in everyday digital media use during the pandemic, mapping digital barriers and strategies to manage digital constraints. The data were analysed using the digital repertoire lens, with close attention paid to how digital routines were shaped by – but not limited to – personal, technological and contextual factors (Hänninen et al. 2021).

However, this chapter also dives deep into the participants' affective digital repertoires as shaped by socio-cultural and technological forces. On top of the interview data, field notes were utilised to jot down initial observations during and after the interviews. The interviews in English were transcribed verbatim. The data were analysed and categorised based on similarities and differences (Saldaña 2011) in using mobile device and online networked platforms. The interview data was coded via NVivo. For this chapter, several quotes based on the interviews are incorporated; research participants' original names have been replaced with pseudonyms.

Staying connected while at home

The onset of the pandemic created multiple forms of disruption among the participants of the study. The implementation of travel restrictions and international border closures and lockdowns forced ageing migrants and their local and transnational networks to 'stay at home'. In Victoria, six lockdowns were imposed after the start of the pandemic in 2020. These 'on and off' lockdowns were implemented to stop the spread of the Covid-19 virus in Victoria, considered the worst affected state in Australia.

Remote interviewing with the participants was conducted in 2020 and 2021. Both interviews with the same participants were deployed during a lockdown. In 2020 the initial interviews focused on the major disruptions that the participants had dealt with, and how the use of mobile technologies and online platforms helped to manage such interruptions. For majority of the participants, the lockdown meant that physical gatherings with families and peers were impossible (see also Popyk and Pustulka 2021).

For instance, I spoke to Edwin, a 66-year-old Indonesian living with his wife in Victoria. Edwin has three children spread across London, New York and Melbourne. On a typical day in Victoria, his daughter and her family would normally visit him and his wife, or sometimes vice versa. However, the lockdown curtailed such visits. Before the pandemic, as Edwin observed, he had been accustomed to visit:

> Twice a week or even more than that. I visit them or they visit us and our grandchild would stay with us. But since the lockdown that's stopped.

During the lockdown, the participants cancelled their overseas travel due to the travel ban. This happened to the Indian couple, Sonny, aged 71, and Lena, aged 68. They moved to Australia in 2012 through the sponsorship of their son, who lived in Melbourne. As much as they wanted to travel back home, they were left with no choice but to stay where they were. Apart from the travel ban, they also took note of the reported cases in India. Sonny explained:

> In March, we wanted to go to India. The thing is the pandemic is not as high so we are going to go, we really want to go. But they cancelled the programme because the pandemic is very high in India.

These examples clearly show the spatial immobilities experienced and managed by the participants.

In 2020 digital communication technologies were utilised by the participants to sustain connections with their local and transnational networks. Studies have shown that connecting locally and transnationally are key cultural practices of ageing migrants in maintaining their cultural identity, social connectedness and belongingness (Baldassar and Wilding 2019) especially during lockdowns and border closures (Popyk and Pustulka 2021; Wang, S. 2024; Wang, W. Y. 2024). I observed that online platforms such as Facebook, messaging applications and videoconferencing tools were utilised (see also Wang, W. Y. 2024). Video-based platforms such as YouTube and Netflix were utilised to access a range of videos. Additionally, some of the participants accessed content from their homelands through streaming services, and sometimes through content on YouTube. Such practices were evident in studies on older migrants' digital practices pre-pandemic (Wilding et al. 2022). These online channels assisted the participants in navigating the lockdown, while remaining connected to their roots and networks.

In 2021 I conducted a follow-up interview with 10 of the 15 participants whom I had interviewed in 2020.[4] The findings showed the continued use of online tools to forge and maintain connections among participants and their online networks. Across the globe the Delta variant was infecting people, causing deaths at a higher rate than had previous variants of Covid-19 (Mishra 2021). A series of lockdowns were imposed in Victoria and most countries continued to implement travel bans and border closures. The participants of the study continued to use digital communication technologies to traverse physical separation from their support networks.

In touch with local networks

Mobile technologies are considered a digital lifeline during a crisis, especially among individuals separated from their support networks (Alencar et al. 2018; Cabalquinto 2022a). The findings of this study revealed that a smartphone bridged the physical separation between participants and their support networks. With ubiquity and accessibility, the smartphone becomes a 'transportal home' – a portable and accessible domestic space where physically separated individuals interact virtually and which embodies a sense of home (Miller et. al. 2021).

A case in point is Victor, a 70-year-old from Hong Kong, and his wife. During the interviews he shared that he and his wife would normally visit his son and the daughter-in-law in their apartment in the city. Sometimes, however, their son and the daughter-in-law would visit them. The family would gather, eat together and engage in conversations. However, the lockdown curtailed these physical visits of family members (see also Popyk and Pustulka 2021). To cope with the enforced separation, Victor, his wife and his son maintained their connections via phone calls. As Victor explained:

> We used to go visit them once a week, you know, as they are in an apartment in the city. But of course, we haven't seen them for many months. So I always talk to them on the phone most of the time.

For Victor and his wife, the calls played a role in initiating a general checking of the family's welfare during turbulent times. These regular calls in fact facilitated a sense of co-presence and enactment of care (Ahlin 2020) for both Victor and his wife.

Some of the participants talked about shifting physical and cultural gatherings to videoconferencing through a messaging application. In normal times gathering in town halls or cultural spaces is vital to ageing migrants in enacting fellowship and belonging (Sokolovsky 2020; Zontini 2015). In the time before the lockdown Maria, a 65-year-old Filipina, would typically join her peers in a migrant organisation for older Filipino-Australian adults. They engaged in conversations, ate, prayed and enjoyed cultural dances and singing. However, these activities were cancelled during lockdown. To re-stage these interrupted activities, Maria participated via online activities during the lockdowns, becoming an active member of a Facebook group chat with her peers in the elderly Filipino community. On top of engaging in conversations and exercises, videoconferencing via Facebook Messenger also enabled 'praying at a distance'. She explained how the group prayed together and chatted on a range of topics:

> The leader, our leader in the handmaids of the Lord, has set up a group for praying the Rosary. So then before 6 o'clock at nighttime, she asks us to join. It's only for Messenger, it's only up to eight people ... We also talk everyday life, or some cooking.

In these examples, we see how familial and community-based ties were sustained by the use of mobile technologies and online platforms. These culturally informed digital practices (Wang, S. 2024; Wang, W. Y. 2024) reflect what scholars have conceptualised as 'digital kinning' – the ways in which culturally-informed connections are enacted through digital connections (Baldassar and Wilding 2019). Notably, the affective experiences of digital connectivity were shaped by social systems and networked affordances. In the case of Victor, the phone calls reflected the notion of generating co-presence driven by performing familial roles (Wilding et al. 2022). Meanwhile, the case of Maria conveyed the ways in which digital technologies are utilised to sustain cultural practices and generative cultural connections (Millard et al. 2018).

In 2021, although the interviews were conducted during a lockdown, the participants also shared their experiences of being physically present with their family members and peers after the lifting of the lockdown for a certain period. For instance, Mohammed, a 77-year-old Syrian, had the chance to have his son and the family in their house where he and his wife prepared exquisite dishes. He described these with relish:

> We have barbecue, a lot of time is barbecue. The barbecue, this is the main dish. We have some, another traditional food here, his kids, they love it. It's about yoghurt cooked with milk, and it's with rice.

Here we see how the cultural practice of getting together to enjoy homeland dishes re-staged cultural connectedness (Hage 1997). On some occasions Maly, a 60-year-old Cambodian active in leading a migrant organisation, also gathered with members of a group. She observed:

> So we started to do cooking again, we invite them to come again, we start to feel a little bit of normal life again.

Indeed, these statements highlighted the desire of older CALD people to gather in physical places and re-stage familial and social gatherings when lockdown was lifted. However, when the participants went back into lockdown, digital devices were relied upon once again.

It is important to note that one of the most significant restrictions during lockdown was the 5km travel limit. This meant that people were allowed to move around within 5km for purchasing basic goods, caring for others, exercising and undertaking work or study if it was not possible to do this at home. As a result, family visits became impossible for some of the participants, so online technologies became the means of sustaining connections. An example of this is Ella, a 65-year-old Taiwanese who lives alone. When the lockdown was lifted Ella was visited by her sister, and she was also able to visit her. However, when lockdown was reimposed, along with the 5km travel limit this became impossible, as her sister lived more than 5km from Ella. The solution, as she explained, was technology: 'We call or we use Zoom to communicate, or WhatsApp to communicate'.

Moreover, praying together online was restored. This was the experience of Mercy who joined the organisation's online prayer group:

> We have a prayer every Thursday with different organisations. [...], different times of these activities.

This practice allowed Mercy to connect with her peers and fulfilled her desire to belong, thus countering feelings of isolation.

Maintaining overseas linkages

A number of studies have illuminated the pivotal role of digital technologies in enabling ageing migrants to remain connected with their overseas networks (Baldassar and Wilding 2019; Baldassar et al. 2020). Transnational connections are fundamental in maintaining cultural and social connectedness (Baldassar and Wilding 2019), but during lockdown the affective textures of transnational connection were shaped by the impacts of global crisis. Exchanges centred on ensuring the safety of family members and peers spread across the world who were dealing with the uneven impact of the pandemic in certain regions and countries. Digital connectivity enables emotive support among older migrants and their distant networks (Popyk and Pustulka 2021).

During the interviews I spoke to Lothika, a 70-year-old Sri Lankan, who had family members dispersed across the globe. Lothika was working in a telecommunication company in Australia and was divorced. For him, a WhatsApp group chat was used to maintain ties with his children in Melbourne, in the US and England. He explained:

> So I chat to them, 'How are you keeping?' 'All right, do you want anything?' Just normal questions the parents will ask. 'Are you happy?' And within a day they always respond to me, because they know I am not working. They don't have to ring me or anything, just send a message asking are you all right. They are quite happy with that.

In the case of Lothika, utilising phatic communication served as an important means to deliver affective and caring messages (Ahlin 2020; Popyk and Pustulka 2021). It also facilitated a sense of co-presence, especially when family members were not available to communicate synchronously due to work commitments.

In a different example, the participants also connected with their overseas kin and peers. Sonny, whom we met earlier in this chapter, used WhatsApp to stay connected with his brother, sisters, their families and other relatives in India. Through their conversations, he was updated regularly regarding their safety. He noted:

> They are following the government instructions, and these are washing hands, social distancing and using gloves and mask. They are following the government, not going to the city, and they are doing tests. And they are staying home. Not going out. Order all the items through online. And getting it from the local supplier.

I then asked Sonny about how he felt towards his connections with his overseas networks. His response was joyful. 'The effect is good. Really good. Most of our friends and relatives in close contact. When you're in India, it is very easy for family, for friends.'

As we have seen, the expressions of gendered care (Kilkey and Merla 2014) were performed and reconfigured by the male participants. Some of the latter were expressing care and concern through digital connectivity, adding insights to previous findings on how male migrants typically perform an authoritative and provider role in transnational familial arrangements (Parreñas 2005). However, it is evident that female participants performed a dual role (Thomas and Lim, 2011) – as both carers and providers – to their overseas networks.

A case in point is that of Mercy. She connected frequently with her brother, his wife and their children in the Philippines. Mercy's family was delivering physical, hands-on care for their 92-year-old mother, while Mercy was sending money to cover her mother's needs. While living away, she cared from afar by videoconferencing her mother on Facebook Messenger, through the assistance of her brother and his family back home. By seeing the condition of her mother via the visuality of videoconferencing, she was reassured that her mother was receiving proper care (see Ahlin 2020). As Mercy commented:

> He's [referring to her brother] now taking care of my mother and of course the wife is very good too, and also my nephews and nieces. Sometimes they send photos as well. One of my nieces sometimes sleeps with her, especially during the time she was sick. So, she was sleeping with my mother in the bedroom and looking after her. Also, some of my nephews were in the house. So, I'm quite happy with what they're doing.

In 2021 I noticed that transnational connections exposed the way in which the Delta variant impacted the living conditions of transnational households. Specifically, the participants began to share their experiences of how their overseas networks navigated the widespread of the virus within their locality. For instance, I interviewed Dorothy, a 70-year-old Filipina, who described how she coped with the issues encountered by her 'left-behind' family members in the Philippines – her sister and a brother. Her family told her that a neighbour had become infected with Covid-19, news that made Dorothy panic. She shared:

> My sister has told us that one of our neighbours, the whole family of those people living in that house, were in quarantine because somebody tested positive to Covid-19. Of course, everybody in this house was quarantined. So it's very scary because it's just a neighbour. I mean a next-door neighbour who tested positive. They might get it too. Too near, too dangerous.

I then asked Dorothy how she addressed the situation. In response she expressed the vital role of the online channel in communicating concern and safety, significantly at the time when her family members back home were also in lockdown. She replied:

> Sometimes I just talk to them because they don't really go out. They themselves are in quarantine. And maybe almost every other day we have to talk to them, encourage them that everything will pass [...] Sometimes we have to tell a story about what we were when we were young, just to lighten the situation [...] We just keep on laughing, and that's a good sign.

As I have shown, digital communication technologies served as vital conduits to enact distant crisis care (Baldassar 2014), especially among older migrants during the pandemic (Popyk and Pustulka 2021; Wang, S. 2024; Wang, W. Y. 2024). However, in the next section I highlight the interferences in digital communication that occurred during the pandemic. They reveal how disproportionate access to resources, competencies and structural inequalities on a local and transnational scale influence the affective dimensions of digital repertoires.

Interference in a digital hub

Echoing many studies on the structural and technological challenges experienced by older migrants in using digital technologies to cope with the pandemic (see Popyk and Pustulka 2021; Wang, S. 2024; Wang, W. Y. 2024), this section illustrates how research participants experienced frustration, discomfort, ambivalence and helplessness in addressing the impacts of the pandemic on their personal and familial lives. This was reflected in how they used digital technologies to enact local and transnational connections. Here I noticed that the digital repertoires were constrained by limited access to resources, digital competencies, the support network's technological environment and the everyday and challenging conditions triggered by the uneven impact of the global health pandemic across individual countries.

Previous studies have exposed how the lack of access to financial resources undermined mediated communication (Leurs 2014; Madianou and Miller 2012; Parreñas 2005). Such findings were echoed in this study. Most participants were already retired, with only three participants still working, two of them full-time and one part-time. Some of the participants relied on their pensions while also receiving financial assistance from family members. During the interview Mohammed, a retired man, explained that he depended on receiving financial support from the government. However, he declared that the amount he and his wife received was insufficient to cover the rent. He said that he had to pay AUD 1,300 per month in rent, while he and his wife each received AUD 784 in official support. As this was not enough, he was also receiving help from his son, something that he felt uncomfortable about.

> We are suffering too much. We have to take some help from our son. I consider it a shame to take from your son.

As a result of this situation, Mohammed was careful about the use of his mobile data at home. When I asked him if he used his mobile internet in his house, he replied 'Not very much because it costs too much'.

It is important to note that the overseas networks of ageing migrants also encountered issues of accessibility, which impacted the participants' affective digital repertoires. This was illustrated by the case of Edwin and his experiences of connecting with his networks in Indonesia. Elsewhere I have shown how electrical and financial issues impacted Edwin's engagement with his 'left-behind' kin in Indonesia in 2020 (Cabalquinto 2022b). Notably, in my follow-up interview with him in 2021, Edwin reiterated how costs impacted on transnational connections:

> The problem is more about the cost. It may sound cheap in Indonesia. But if you don't have an income then it's not cheap.

I asked Edwin to clarify this further. He referred to the case of a younger sister whom he was supporting:

> My younger sister, she had these ongoing problems with her phone, the laptop and with being able to have internet access on a reliable basis. So that's the challenge that she has and a challenge that we face.

Edwin also explained that the two of them were not videoconferencing frequently because 'it would cost too much bandwidth'.

Studies have shown the role of support networks in assisting older adults (Hänninen et al. 2020) and older migrants (Worrell 2021) in using modern communication technologies. During the lockdown computer and mobile phone shops were closed; tech support was only delivered via a phone call. In addition, support networks were not allowed to undertake in-person assistance with ageing migrants (Cabalquinto 2022b). As a result some of the participants struggled with solving technical issues without in-person assistance. This was the experience of Mercy.

> I bought a laptop just two months ago. The problem with this device is that I cannot just get into my training thing because of the pop-ups. So, when the pop-ups come, and even if you follow the Google thing like how to get rid of the pop-ups and all these things, it doesn't work. Sometimes it doesn't work, especially as the laptop's security feature is quite high. So yeah, I have difficulty with that. I tried a few times ringing the shop but they don't answer. They are too busy. I went to the shop and you have to line up with this Covid-19 protocol; now they're all closed.

I have highlighted elsewhere that ambivalence is generated through the disproportionate living conditions of migrants and their distant networks as visibilised and negotiated through digital practices (Cabalquinto 2022a). For this study, due to the uneven impact of the pandemic across different countries, the participants expressed their feelings of ambivalence and helplessness through digital media use. For instance, Maria stayed connected with her family members and friends in the Philippines via Facebook Messenger. Through the online platform, she knew their welfare, 'They're all right, thank God, and his family.' However, she also felt worried about her native country. As she lamented:

> I feel worried about what will happen to the Philippines, you know. I am worried about the people who are dying there. People that are getting hungry, dying through not eating. And our President [referring to inadequate pandemic responses]. That's what I'm worried about, and whenever I hear something bad about our country.

In 2020 the participants did not report anyone in their families having been infected by Covid-19. However, in several follow-up interviews in 2021 two participants shared their experiences as they went through the ordeal of infection and death among distant family members. It was at this time that the Delta variant became a dominant strain across the world. A case in point was Edwin, originally from Indonesia, who explained:

> My nephew's son, my grandnephew, has had Covid as a baby. So very young, so at two months old he had Covid, but has now recovered, which is good. So that's the closest one in my family, but none of my siblings luckily, nor my siblings' children. So that one there with the baby with Covid, the parents didn't get Covid but it came from their hired help. [...] So there is a lot of my cousins have got Covid and sadly some may have also passed away because of Covid.

Edwin knew about the cases through their family group chat in WhatsApp. I then asked him about how he dealt with the situation. He said that he provided emotional and financial support, a practice symbolising 'being there' for overseas networks (Baldassar et al. 2007; Cabalquinto 2022a). As he said:

> I would provide financial support in terms of just contribution for people who are needing financial help, especially to those who have a family member who has passed away. We would send them some financial contribution for that, and the whole family group would do that.

Conclusion

Mobile technologies are considered a digital lifeline during a crisis, especially among individuals separated from their support networks (Alencar et al. 2018; Cabalquinto 2022a). This chapter has showcased the digital repertoires of ageing migrants during a series of lockdowns in 2020 and 2021 in Victoria, Australia. To begin with, it presents how ageing migrants used a diverse range of technological apparatus to forge and maintain ties within and beyond borders, contributing to a sense of mediated co-presence. The networked connectivity of online platforms has facilitated interactions between ageing migrants and their local and transnational networks, thus activating cultural connectedness and positive feelings. Indeed, the participants' physical mobility – gathering and interacting with family and friends – was put on hold during the pandemic, making digital connectivity the only way to traverse physical separation and enable imagined co-presence.

It is crucial to point out that the study has shown how digital repertoires were impacted by an individual's limited access and competencies – as well as the varied social and technological conditions of local and transnational linkages. We see how constraints in digital media use produced discomfort, sadness and feelings of helplessness; not everyone was afforded with stable and comfortable connections. For those who had adequate digital access and competencies, paired with accessible support networks, staying at home while digitally connected offered a comforting and secure experience. Yet others, by contrast, experienced digital repertoires as a source of pain and ambivalence, with everyday digital routines being shaped by one's already constrained personal, social and technological situation.

A digital repertoire perspective has been useful in unravelling both the benefits and limits of digital media use among older migrants during the Covid-19 pandemic. It highlights the heterogeneity of everyday digital media use (Hänninen et al. 2021a) as shaped by social, cultural, situational and environmental, even infrastructural factors. In such a case it provides a critical entry point for future research to reflect on the impacts of digitalisation among vulnerable groups, by focusing upon the intersecting influences of household obligations, technological affordances and even broader economic and political conditions, such as the ways in which governments respond to a crisis and assist individuals, families and even communities. In doing so, critical approaches do not only unlock the positive affective states of older migrant and their implications for ageing well. Such interventions can also locate and redress unsettling experiences of digital connectivity, enacted both locally and transnationally, especially at a time of a crisis in which anxiety and stress are amplified.

Acknowledgements

This project is supported by The Alfred Deakin Institute for Citizenship and Globalisation Research Pilot Grant [Grant Number 618575789].

Notes

1 This term has been used in Australia to refer to individuals with non-speaking English backgrounds. CALD people can be resident in Australia on either a temporary or permanent status.

2 The study acknowledges that the recruitment of participants via digital channels failed to include those who did not have material access to digital technologies, as well as those with limited technological competencies, to access information online (Cabalquinto and Ahlin 2023).
3 For a detailed account on the author's reflection and positionality in deploying the research methods, see Cabalquinto and Ahlin, 2023.
4 Only 10 were engaged in a follow-up interview. The remaining five did not opt to participate in a follow-up interview as stated in the consent form.

References

Ahlin, T. 2020. 'Frequent callers: "Good care" with ICTs in Indian transnational families', *Medical Anthropology,* 39(1): 69–82. https://doi.org/10.1080/01459740.2018.1532424.

Ahmed, S. 2004. 'Affective economies', *Social Text* 79(22): 117–40. https://muse.jhu.edu/article/55780/summary.

Ahmed, S. 2014. *The Cultural Politics of Emotion* (second edition). Edinburgh: Edinburgh University Press.

Alencar, A., K. Kondova and W. Ribbens. 2018. 'The smartphone as a lifeline: An exploration of refugees' use of mobile communication technologies during their flight', *Media, Culture & Society* 41(6): 828–44. https://doi.org/10.1177/0163443718813486.

Alinejad, D. 2019. 'Careful co-presence: The transnational mediation of emotional intimacy', *Social Media + Society* 5(2): 1–11. https://doi.org/10.1177/2056305119854222.

Ash, J. 2015. 'Technology and affect: Towards a theory of inorganically organised objects', *Emotion, Space and Society* 14: 84–90. https://doi.org/10.1016/j.emospa.2013.12.017.

Australian Bureau of Statistics. 2021. *Migration, Australia*. Accessed 2 February 2022. https://www.abs.gov.au/statistics/people/population/migration-australia/latest-release.

Baldassar, L. 2008. 'Missing kin and longing to be together: Emotions and the construction of co-presence in transnational relationships', *Journal of Intercultural Studies* 29(3): 247–66. https://doi.org/10.1080/07256860802169196.

Baldassar, L. 2014. 'Too sick to move: Distant "crisis" care in transnational families', *Revue Internationale de Sociologie* 24(3): 391–405. https://doi.org/10.1080/03906701.2014.954328.

Baldassar, L., C. Baldock and R. Wilding. 2007. *Families Caring Across Borders: Migration, ageing and transnational caregiving*. Houndmills, Basingstoke: Palgrave Macmillan.

Baldassar, L. and L. Merla. 2014. 'Introduction: Transnational family caregiving through the lens of circulation'. In *Transnational Families, Migration and the Circulation of Care: Understanding mobility and absence in family life,* edited by L. Baldassar and L. Merla, 3–24. New York: Routledge.

Baldassar, L. and R. Wilding. 2019. 'Migration, ageing, and digital kinning: The role of distant care support networks in experiences of ageing well', *The Gerontological Society of America* 60(2): 1–9. https://doi.org/10.1093/geront/gnz156.

Baldassar, L., R. Wilding and S. Worrell. 2020. 'Elderly migrants, digital kinning and digital home making across time and distance'. In *Ways of Home Making in Care for Later Life*, edited by B. Pasveer, O. Synnes and I. Moser, 41–62. Los Angeles, CA; Singapore: Palgrave Macmillan.

Boccagni, P. 2022. 'Transnational migration and homemaking'. In *Handbook on Transnationalism*, edited by B. S. A. Yeoh and F. L. Collins, 141–54. Cheltenham, UK; Northampton, MA: Edward Elgar Publishing. https://doi.org/10.4337/9781789904017.00016.

Cabalquinto, E. C. 2020. 'Researching transnational family life in the age of mobile media'. In *Handbook on Methods and Applications for Mobilities Research*, edited by M. Büscher, M. Freudendal-Pedersen and S. Kesselring, 251–62. Cheltenham, UK: Edward Elgar Publishing. https://doi.org/10.4337/9781788115469.00032.

Cabalquinto, E. C. 2022a. *(Im)mobile Homes: Family life at a distance in the age of mobile media.* Oxford: Oxford University Press. https://doi.org/10.1093/oso/9780197524831.001.0001.

Cabalquinto, E. C. 2022b. '"Without technology we'd be very stuck": Ageing migrants' differential (im)mobile practices during a lockdown', *Media International Australia,* 188(1): 3–17. https://doi.org/10.1177/1329878X221095582.

Cabalquinto, E. C. 2022c. '"Come on, put Viber, we can drink coffee together": Performing (im) mobile intimacy in turbulent times among aging migrants', *Communication Culture and Critique* 15(2): 244–60. doi:https://doi.org/10.1093/ccc/tcac011.
Cabalquinto, E. and T. Ahlin. 2023. 'Researching (im)mobile lives during a lockdown: Reconceptualising remote interviews as field events', *International Journal of Cultural Studies* 26(5): 802–21. https://doi.org/10.1177/13678779231157428.
Cuban, S. 2018. 'Rescue chains and care talk among immigrants and their "left-behind" parents'. In *Connecting Families? Information and Communication Technologies, generations and the lifecourse*, edited by B. Barbosa Neves and C. Casimiro, pp. 219–35. Bristol: Bristol University Press.
Fernández-Ardèvol, M. 2020. 'Older people go mobile'. In *The Oxford Handbook of Mobile Communication and Society*, edited by R. Ling, L. Fortunati, G. Goggin, S. S. Lim and Y. Li. pp. 187–199. Oxford: Oxford University Press.
Garattini, C. and D. Prendergast. 2017. 'Critical reflections on ageing and technology in the twenty-first century'. In *Aging and the Digital Life Course*, edited by D. Prendergast and C. Garattini, 1–15. New York: Berghahn Books.
Hage, G. 1997. 'At home in the entrails of the west: Multiculturalism, ethnic food and migrant home-building'. In *In Home/World: Space, community and marginality in Sydney's West*, edited by H. Grace, G. Hage, L. Johnson, J. Langsworth and M. Symonds, 99–153. Sydney: Pluto Press.
Hänninen, R., L. Pajula, V. Korpela and S. Taipale. 2021a. 'Individual and shared digital repertoires – older adults managing digital services', *Information, Communication & Society* 26(3): 1–16. https://doi.org/10.1080/1369118X.2021.1954976.
Hänninen, R., S. Taipale and A. Korhonen. 2021b. 'Refamilisation in the broadband society: The effects of ICTs on family solidarity in Finland', *Journal of Family Studies* 27(2): 161–77. https://doi.org/10.1080/13229400.2018.1515101.
Hänninen, R., S. Taipale and R. Luostari. 2020. 'Exploring heterogeneous ICT use among older adults: The warm experts' perspective', *New Media & Society* 23(6): 1584–1601. https://doi.org/10.1177/1461444820917353.
Hargittai, E., A. M. Piper and M. R. Morris. 2019. 'From internet access to internet skills: Digital inequality among older adults', *Universal Access in the Information Society* 18(4): 881–90. https://doi.org/10.1007/s10209-018-0617-5.
Helsper, E. 2021. *The Digital Disconnect: The social causes and consequences of digital inequalities*. London: SAGE.
Hjorth, L., S. Pink, S. and H. Horst. 2018. 'Being at home with privacy: Privacy and mundane intimacy through same-sex locative media practices', *International Journal of Communication* 12: 1209–27. https://ijoc.org/index.php/ijoc/article/view/7050/2295.
Holmes, M. and R. Wilding. 2019. 'Intimacies at a distance: An introduction', *Emotion, Space and Society* 32: 100590. https://doi.org/10.1016/j.emospa.2019.100590.
Horst, H. 2011. 'Reclaiming place: The architecture of home, family and migration', *Anthropologica* 53(1): 29–39. https://www.jstor.org/stable/41475727.
Hutchby, I. 2014. 'Communicative affordances and participation frameworks in mediated interaction', *Journal of Pragmatics*. 72: 86–9. https://doi.org/10.1016/j.pragma.2014.08.012.
Ito, M. 2005. 'Intimate visual co-presence'. [Online article]. http://www.itofisher.com/mito/archives/ito.ubicomp05.pdf.
Kilkey, M. and L. Merla. 2014. 'Situating transnational families' care-giving arrangements: The role of institutional contexts', *Global Networks* 14(2): 210–29. https://doi.org/10.1111/glob.12034.
Lamb, S. 2020. 'Old age homes, love and other new cultures of aging in middle-class India'. In *The Cultural Context of Aging: Worldwide perspectives*, edited by J. Sokolovsky, 466–91. Santa Barbara, CA: Praeger.
Lasén, A. 2004. 'Affective technologies: Emotions and mobile phones' [Online article]. Accessed 14 June 2022. https://www.receiver.vodafone.com.
Leurs, K. 2014. 'The politics of transnational affective capital: Digital connectivity among young Somalis stranded in Ethiopia', *Crossings: Journal of Migration & Culture* 5(1): 87–104. https://doi.org/10.1386/cjmc.5.1.87_1.
Licoppe, C. 2004. '"Connected" presence: The emergence of a new repertoire for managing social relationships in a changing communication technoscape', *Environment and Planning D: Society and Space* 22(1): 135–56. https://doi.org/10.1068/d323t.

Lim, S. S. 2016. 'Asymmetries in Asian families' domestication of mobile communication'. In S. S. Lim (Ed.), *Mobile Communication and the Family: Asian experiences in technology domestication*, edited by S. S. Lim, 1–12. London: Springer.
Ling, R. S. 2008. *New Tech, New Ties: How mobile communication is reshaping social cohesion*. Cambridge, MA: MIT Press.
Lünenborg, M. and T. Maier. 2018. 'The turn to affect and emotion in media studies', *Media and Communication* 6(3): 1–4. https://doi.org/10.17645/mac.v6i3.1732.
Madianou, M. 2016. 'Ambient co-presence: Transnational family practices in polymedia environments', *Global Networks* 16(2): 183–201. https://doi.org/10.1111/glob.12105.
Madianou, M. and D. Miller. 2012. *Migration and New Media: Transnational families and polymedia*. Abingdon, Oxon: Routledge.
Millard, A., L. Baldassar and R. Wilding. 2018. 'The significance of digital citizenship in the well-being of older migrants', *Public Health* 158: 144–8. https://doi.org/10.1016/j.puhe.2018.03.005.
Miller, D., L. Abed Rabho, P. Awondo, M. de Vries, M. Duque, P. Garvey, L. Haapio-Kirk, C. Hawkins, A. Otaegui, S. Walton and X. Wang. 2021. *The Global Smartphone: Beyond a youth technology*. London: UCL Press. https://doi.org/10.14324/111.9781787359611.
Mishra, S. 2021. 'The Delta variant is spreading fast, especially where vaccination rates are low', *National Geographic*, 9 July 2021. Accessed 1 February 2022. https://www.nationalgeographic.com/science/article/the-delta-variant-is-serious-heres-why-its-on-the-rise.
Moore, R. C. and J. T. Hancock. 2020. 'Older adults, social technologies and the coronavirus pandemic: Challenges, strengths and strategies for support', *Social Media + Society* 6(3). https://doi.org/10.1177/2056305120948162.
Nedelcu, M. 2012. 'Migrants' new transnational habitus: Rethinking migration through a cosmopolitan lens in the digital age', *Journal of Ethnic & Migration Studies* 38(9): 1339–56. https://doi.org/10.1080/1369183X.2012.698203.
Nedelcu, M. and M. Wyss. 2016. '"Doing family" through ICT-mediated ordinary co-presence: Transnational communication practices of Romanian migrants in Switzerland', *Global Networks* 16(2): 202–18. https://doi.org/10.1111/glob.12110.
Parreñas, R. S. 2005. 'Long distance intimacy: Class, gender and intergenerational relations between mothers and children in Filipino transnational families', *Global Networks* 5(4): 317–36. https://doi.org/10.1111/j.1471-0374.2005.00122.x.
Patterson, J. and K. Leurs. 2020. 'Transnational digital intimacy practices: Paradoxes of transnational connectivity and home-making among young adult expatriates in Amsterdam, the Netherlands', *Global Perspectives* 1(1): 1–17. https://doi.org/10.1525/gp.2020.13346.
Peile, C. 2018. 'Transnational family communication during an economic crisis: Personal media repertoires of Moroccans and Ecuadorians in Spain'. In *Diaspora and Media in Europe: Migration, identity and integration*, edited by K. H. Karim and A. Al-Rawi, 127–46. Cham, Switzerland: Springer International Publishing AG.
Popyk, A. and P. Pustułka. 2021. 'Transnational communication between children and grandparents during the Covid-19 lockdown: The case of migrant children in Poland', *Journal of Family Communication* 21(3): 223–37. https://doi.org/10.1080/15267431.2021.1929994.
Richardson, I. and R. Wilken. 2012. 'Parerga of the third screen: Mobile media, place and presence'. In *Mobile Technology and Place*, edited by G. Goggin and R. Wilken, 198–212. New York: Routledge.
Saldaña, J. 2011. *Fundamentals of Qualitative Research*. Oxford: Oxford University Press.
Sampaio, D. 2020. 'Caring by silence: How (un)documented Brazilian migrants enact silence as a care practice for ageing parents', *Journal of Intergenerational Relationships* 18(3): 281–300. https://doi.org/10.1080/15350770.2020.1787038.
Sinanan, J. 2019. 'Visualising intimacies: The circulation of digital images in the Trinidadian context', *Emotion, Space and Society* 31: 93–101. https://doi.org/10.1016/j.emospa.2019.04.003.
Sokolovsky, J. 2020. 'Introduction: A 21st-century global perspective on ageing and human maturity in cultural context'. In *The Cultural Context of Ageing: Worldwide perspectives*, edited by J. Sokolovsky, xvii–liii. Santa Barbara, CA: Praeger.
Taipale, S. 2019. *Intergenerational Connections in Digital Families*. Cham, Switzerland: Springer International Publishing.
Tettegah, S. Y. 2016. *Emotions, Technology and Social Media*. London: Academic Press.

Thomas, M. and S. S. Lim. 2011. 'On maids and mobile phones: ICT use by female migrant workers in Singapore and its policy implications'. In J. Katz (Ed.), *Mobile Communication and Social Policy*, edited by J. Katz, 175–90. New Jersey: Transaction.
Twigt, M. 2022. *Mediated Lives: Waiting and hope among Iraqi Refugees in Jordan*. New York: Berghahn.
Vincent, J. and L. Fortunati. 2009. *Electronic Emotion: The mediation of emotion via Information and Communication Technologies*. New York: Peter Lang.
Wang, S. 2024. 'Social ties of elderly migrants during Covid-19: The Chinese in Paris'. In *Chinese in France amid the Covid-19 Pandemic: Daily lives, racial struggles and transnational citizenship of migrants and descendants*, edited by Simeng Wang, 177–95. The Netherlands: Brill.
Wang, W. Y. 2024. 'Scaling the "ageing migrant body" in the digital era: A case of older Chinese migrants in Australia during the Covid-19 pandemic', *Journal of Intercultural Studies* 1–19. https://doi.org/10.1080/07256868.2024.2307952.
Wilding, R. 2006. '"Virtual" intimacies? Families communicating across transnational contexts', *Global Networks* 6(2): 125–42. https://doi.org/10.1111/j.1471-0374.2006.00137.x.
Wilding, R., S. Gamage, S. Worrell and L. Baldassar. 2022. 'Practices of "digital homing" and gendered reproduction among older Sinhalese and Karen migrants in Australia', *Journal of Immigrant & Refugee Studies* 20(2): 220–32. https://doi.org/10.1080/15562948.2022.2046895.
Wise, A. and S. Velayutham. 2017. 'Transnational affect and emotion in migration research', *International Journal of Sociology* 47(2): 116–30. https://doi.org/10.1080/00207659.2017.1300468.
Witteborn, S. 2014. 'Forced migrants, emotive practice and digital heterotopia', *Crossings* 5(1): 73–85. https://doi.org/10.1386/cjmc.5.1.73_1.
Worrell, S. 2021. 'From language brokering to digital brokering: Refugee settlement in a smartphone age', *Social Media + Society* 7(2): 1–11. https://doi.org/10.1177/20563051211012365.
Zontini, E. 2015. 'Growing old in a transnational social field: Belonging, mobility and identity among Italian migrants', *Ethnic and Racial Studies* 38(2): 326–41. https://doi.org/10.1080/07256868.2024.230795210.1080/01419870.2014.885543.

5
Digital repertoires of care in Japan: a participatory visual approach

Laura Haapio-Kirk

Introduction

This chapter examines how digital repertoires among older adults in Japan are developed in correspondence with values of care. Goodwin (2015) argues that care is realised through embodied acts that communicate affect, attention, empathy and co-presence. I extend these ideas of embodiment and proximity as central to care by exploring how digital visual communication can create an aesthetics of care that draws on embodied expression, using a participatory visual methodology to uncover ethnographic insights. If digital repertoires are developed in tandem with relationships of care, then in order to understand the role of technologies in people's daily lives we must as researchers find ways to reach private moments and spaces in which care is realised. This chapter explores collaborative comics and film-making as two such approaches.

I present evidence gathered both through in-person object elicitation interviews and methodological experiments with co-produced comics and videos, showing how digital repertoires are shaped in relation to meaningful aspects of people's lives. These participatory methods not only reveal insights that would have otherwise remained invisible, but also make visible embodied and spatially contextual practices that would be difficult to convey in words alone.[1] By developing a graphic and filmic approach to studying digital repertoires, I am drawing upon scholarship in visual anthropology that urges a diversification from the centrality of text (Banks and Zeitlyn, 2015; Cox et al. 2016; Pink 2012; Stoller 1997). I assert that graphic multimodal approaches make it possible to convey multisensorial and embodied experiences in a medium that can be easily reproduced on the page but is not limited to text. In so doing I take Dattatreyan and Marrero-Guillamón's (2019, 220) definition of multimodal work as 'multisensorial rather than text based, performative rather than representational, and inventive rather than descriptive'.

Rather than focus on digital divides or the ways in which older adults become digitally skilled as they adopt the smartphone, this chapter demonstrates how personality, interests and cultural contexts shape emerging digital repertoires. I find that these repertoires, spanning a range of digital technologies and platforms, are being used to mitigate some sense of vulnerability and dependency on others as ageing persons, even as they simultaneously embed people in relationships of care. The case studies highlight the diversity of experiences among older adults, demonstrating how individual motivations and concerns shape digital repertoires that are not dependent on age categorisation.

From *garakei* to smartphones

The data I draw on in this chapter comprises evidence gathered during 16 months of in-person ethnographic fieldwork, conducted in Japan between February 2018 and May 2019 and supplemented by continued remote fieldwork from May 2019 to March 2022. This was a time during

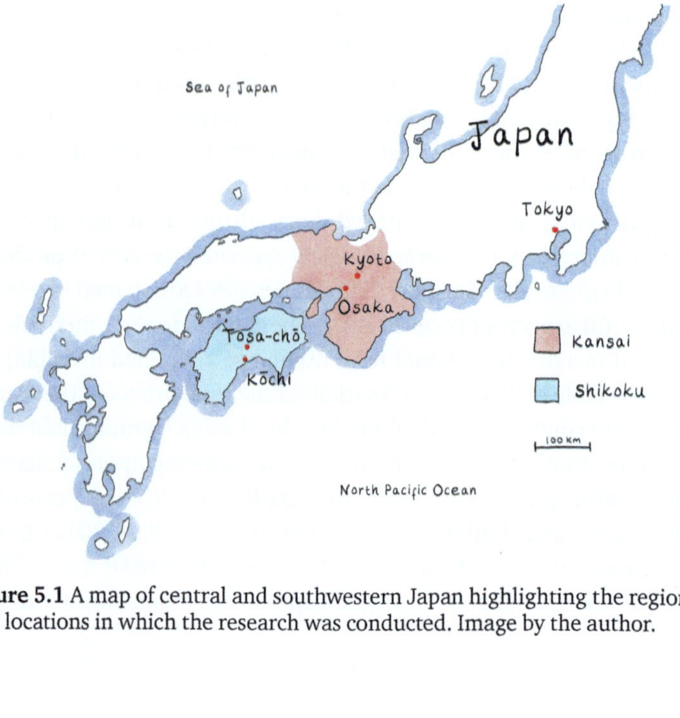

Figure 5.1 A map of central and southwestern Japan highlighting the regions and locations in which the research was conducted. Image by the author.

which many adults were switching from their old feature phones (known as '*garakei*') to smartphones. Fieldwork was conducted primarily in two field sites; the Kansai region of Western Japan, involving participants located in the cities of Osaka and Kyoto, and in Shikoku, southwestern Japan, in a rural town called Tosa-chō in Kōchi Prefecture (Fig. 5.1).

First, let me provide a brief overview of Japan's mobile phone landscape among older adults at the time of research. Japan initially led innovation in mobile communication; mobile phones in Japan have been defined by their provision of internet access – in other words, they have been 'smart' – for a lot longer than elsewhere.[2] Internet-connected feature phones (Fig. 5.2) became popular a decade before the advent of the smartphone (Ito 2005). The demand for mobile phones had become particularly strong among young people, building on the popularity of pagers for casual communication among young people at the time (Ito 2005). '*Keitai*' (mobile) culture was thus shaped by the creative usage by young people, marking the birth of the mobile society in Japan (Okada 2005). However, my research indicates that smartphones are now reaching all demographics of society,[3] not through top-down efforts, but typically because middle-aged children and relatives are giving smartphones to older adults to improve how they care for them.

While the popularity of *garakei* was initially driven by adoption from teenagers who shifted the association of the mobile phone with business to teen street culture (Ito 2005), my research found that in the smartphone era *garakei* have become synonymous with older users of mobile phones. My older research participants particularly valued their hard-wearing design, long battery life and non-touch screen interface. They also liked the simplicity of the *garakei*, with voice calling, messaging and step counter being among the top three uses for their phones. However, during the period of my ethnography on which this chapter is based (between 2018 and 2019), many older people were switching to smartphones. This made it the perfect time not only to assess the impact of the smartphone on their lives, but also to understand why some were reluctant to give up their *garakei,* as shown in the case studies below. Many of my participants had both a *garakei* and a smartphone, while others had only a *garakei* and a few people only had landlines. Rather than this being determined by age, I found that it was more influenced by gender. More of my women participants than men were switching to smartphones, largely for social reasons.

Smartphone ownership among seniors has rapidly increased over the past few years, as the Japanese government put pressure on telecommunication companies to reduce the price of packages, thus

Figure 5.2 A man holding his 'feature phone', locally known as '*garakei*'. Photo by the author.

bringing them into line with other countries. Shortly after I left Japan in the summer of 2019, friends reported that long queues were forming outside mobile phone shops as alluring promotions were applied to smartphones. Several of my participants changed to smartphones at this time, including Maiko-san, a woman in her sixties who had been adamant that she would never get one. Her story is presented below.

The Covid-19 pandemic has further accelerated the uptake of smartphones among older demographics. As of 2021, the smartphone ownership rate among Japanese older adults was 80 per cent for those in their sixties and 62 per cent for those in their seventies, an increase of 11 per cent and 14 per cent respectively since 2020.[4] 'Family recommendations' tended to be the reason for the switch, given especially by women. My ethnography situated the financial burden of smartphone adoption within a range of other factors that influence whether someone switches to a smartphone or not. These include an individual's openness towards trying new technology, their attachment to a network of contacts perceived to be contained in the *garakei* and sometimes a simple matter of not wanting to let go of a seldom-used phone number.

From my observations of, and discussions with, older adults about their use of *garakei* and smartphones, it was clear that while *garakei* are internet-enabled, the adoption of the smartphone afforded many more capabilities than were possible on their older devices. Many of these affordances were visual in nature. For example, while *garakei* typically had integral cameras, the resolution was low and there could be difficulties in sending and receiving photos between smartphones and *garakei*. People who made the switch found themselves able to participate in more visual forms of communication through, for example, the sharing of photographs. They were also able to use the communication application LINE, which represented a radical change in their mobile phone usage and communication possibilities.

LINE was the most used app among my research participants in both Kyoto and rural Kōchi. I did not come across anyone with a smartphone who did not use the application; some users even referred to their smartphone as 'my LINE'. I found that an important aspect of this key platform were the affordances of LINE that facilitate visual communication in the form of photos, emojis and stickers. Visual messaging became a particularly distinctive aspect of the digital repertoires of older, particularly female, participants as a mediator of care.

While young people may have originally shaped mobile culture in Japan (Okada 2005), older people now make up a large proportion of mobile phone users. They are, in their own way, shaping smartphone culture, even as mobile phones themselves are also adapting to older people. Popular with several of my older interlocutors are the so-called 'easy (*raku raku*) smartphones'. Developed by a number of Japanese technology companies including SoftBank (Murai 2014), Fujitsu and NTT DOCOMO (Byford 2012), they are equipped with features such as magnifying software and an easily navigable interface with large icons and text. Both the participants featured below used such 'easy' smartphones, but they had quite different trajectories towards acquiring them.

Sato-san

In a traditional wooden house (*machiya*) on a quiet street in central Kyoto lives Sato-san, a sprightly *ikebana* (flower arranging) master whose life is intertwined with the evolution of modern Japan. Born in 1928, she was nearly 90 years old when I first met her in her home one afternoon,[5] having been introduced by Maiko-san, one of her students who will be discussed below. Sato-san described how, having never married, she had lived in the same house her whole life, apart from in her teenage years during the Second World War when she stayed with relatives outside Kyoto. The house had changed little since it was built by her father in the Meiji era (between 1868 to 1912), except for a new roof and floor.

In the centre of the main room was a big table on which Sato-san made her floral arrangements. She supplied these to selected shrines and businesses, such as a local *wagashi* (traditional Japanese confectionery) shop. Her home was also where she taught her students, working four days per week. She took photos of her arrangements and her niece's son helped her to uploaded them to her blog, through which some of her students found her. As we talked Sato-san shared glimpses of her daily life which, despite the traditional surroundings, was intricately entwined with digital technologies – primarily her 'easy' (*rakuraku*) smartphone designed for older adults.

Sato-san recounted the story of her unexpected journey into the world of smartphones two years before we met. 'Well, I always had a *garakei*,' she chuckled, reminiscing about her trusty old flip phone. 'And then, my niece, she's 21 years younger than me, she was changing hers to a smartphone and said, "right, we're changing yours too!"' She laughed, clearly amused by the memory. 'So I was more forced to than by choice!' Sato-san had no children of her own, so her niece had taken on the role that I found might normally be played by a son or daughter, caring for her aunt by making sure she was digitally connected.

While the switch to a smartphone was not her choice, Sato-san soon found that she came to rely on the new device. 'But it's really convenient! I can look up the weather and see if it's going to rain in my area … I just type in [place name]'. Checking the weather was of particular importance to Sato-san because of her physical limitations. 'You know I have a bad leg? I can't hold an umbrella! I have to walk like this,' she gestured, demonstrating her bent-over shuffle. 'So I check how much it's raining.' If rain was forecast she would stay at home or catch a bus, looking up the timetable on her phone. The phone further served her mobility by providing health information via Google searches about how to improve her swollen legs.

One of the central motivations shaping Sato-san's digital repertoire was her concern to care for her health. She was committed to keeping her pedometer app happy with regular walking and she also used various brain training apps daily. 'I use my smartphone every day. I do brain training! And "One-word English"! Yes, every day,' she exclaimed with a gleam in her eye. However, her love for learning began years before and was motivated by more than maintaining brain health in later life. Sato-san recounted the disruption to her English language education during her high school years after the Second World War, when English was considered the language of the enemy. Unlike her sister, who was eight years older than her, Sato-san never felt that she had received the education she wanted.

The war broke out just when Sato-san entered high school. She persevered by subsequently attending classes at monasteries and became one of the first people to take the new English proficiency test (*eiken*) when it made its debut after the war. Recalling the test with a chuckle, she reminisced, 'There was a microphone, and oh, it was such bad quality! It's not like today where things are good quality. So the voices broke a lot, and it was difficult to do the listening test'. In the same breath as talking about the poor-quality microphone and speaker, she recalled how at this time everyone wore *geta* (traditional Japanese wooden sandals) and not the Western-style shoes that are now the norm. Sato-san's recollections of the change in footwear parallelled the broader transformations she had witnessed during Japan's ascent as a leader in technology in the second half of the twentieth century. As she reflected on this evolution, it served as a poignant reminder of the rapid pace of change that had characterised Japan's modernisation journey – a journey that she cared about and felt part of through her own adoption of digital technologies and general attitude towards learning.

In her seventies, Sato-san once again enrolled in English classes. However, her worsening eyesight made it difficult to see the board and she also felt out-of-step with the conversations of the other students, all aged in their twenties. She did not want to feel like a burden, so eventually she left. Despite these challenges, Sato-san's passion for learning never waned. Now, using her smartphone, she was able to learn new words every day. It was not only English that Sato-san was curious about, but new Japanese words too. She explained:

> On TV too, there are so many words that are difficult these days, especially in politics. So I always look it up on this [holding her smartphone]! Yes! It's so much faster than looking it up [in a book]! I always used to wonder 'what are they talking about?'

The way that young people spoke was different these days, she felt, and sometimes hard to understand. Their greetings were also shorter and less varied, something she felt was a loss. At the heart of Sato-san's character was a profound interest in comprehending the world around her, reflecting her genuine care and empathy for those within her sphere of influence. This inclination transferred to the way in which she communicated via her smartphone.

Sato-san's care for others shone through in her adaptability and thoughtful approach to communication. She eagerly talked about the convenience of using LINE to stay in touch with her students, recognising the challenge of coordinating with individuals busy with their own work schedules. As she proudly mentioned her growing list of contacts on LINE (it had then reached 101), it became evident that Sato-san's commitment to fostering connections went hand in hand with her willingness to adapt her communication style to suit best the needs of those she cared about. Her students would email her photos of their *ikebana* work and she would write long replies, even though she found it cumbersome to type on her smartphone keyboard. Her LINE messages, in contrast, were shorter, but peppered with emojis and stickers showing friendly characters, often surrounded by hearts. In this way her messaging style mirrored how she viewed communication in Japan to be changing more broadly, with people preferring to use more concise language. Visual communication was thus an acceptable shorthand for Sato-san, taking on the role of affectionate greetings without the need for cumbersome typing.

The way that people spoke was not the only thing that had altered in Sato-san's lifetime. The area around her home was also changing fast. New hotels were springing up everywhere and tourists were a common sight in the small narrow streets. Sometimes Sato-san would try to help a lost-looking tourist, using her smartphone to communicate with people of different languages. 'I can just insert here what I want to say and up it pops in English! So convenient,' she exclaimed, referring to a translation app. As she demonstrated the features of her 'easy smartphone', changing camera angles and showing how she would take a selfie, Sato-san's enthusiastic and curious personality emerged as central to her adaptability to new technologies. As she herself observed:

> I've had that since I was very young. So curious, such a curious mind! I would always ask why?? And every time I did, they would answer 'because it is so!' [laughs]

Sato-san has been able to adapt to the changing times with enthusiasm because of her openness to trying new things. She felt this singled her out from others her age, or even from those younger than herself.

Maiko-san

Maiko-san, a woman in her late sixties, is one of Sato-san's students. She goes every two weeks to *ikebana* classes and has developed a friendship with her teacher, whom she considers an inspiration. When I was doing my fieldwork Maiko-san staunchly refused to switch from a *garakei* to a smartphone despite having friends around her who were making the change. She wore bifocals and thought it would be awkward to try to read messages on a small screen. With a computer at home to access the internet and send emails, she felt no need to depart from her flip phone. She also valued the sense of privacy and detachment from social pressures that it gave her. She was wary of being always available and worried by the added pressure that a smartphone would bring:

> I don't want to be controlled by it. It's becoming like you have to answer right away, like you have to email back right away. Before, an answering machine was enough! ... It's too much.

Whereas Sato-san, 20 years older than Maiko-san, exuded enthusiasm for new technologies, Maiko-san was less convinced. Sato-san's adventurous spirit and openness to new experiences defy stereotypes about older adults' reluctance to adopt new technologies, demonstrating an inherent resilience and curiosity that transcends age. The two women illustrate the heterogeneity of older adults, with digital repertoires shaped by their own personal histories and individual personalities rather than their respective ages. Of course, that is not to deny that digital skills were less prevalent generally among older cohorts, as Sato-san herself appreciated: 'Everyone finds it [the smartphone] hard, but for me... [smiles]'.

By presenting the case of Sato-san, an outlier among her peers, my intention is not to diminish the real difficulties and obstacles facing older adults in the adoption of digital technologies. Rather, I wish to show how these are not determined by age alone. Sato-san's open approach had been cultivated over a lifetime, and was not limited to her digital repertoires. She explained:

> I think it's so fun to try new things. Like parasailing! The only thing I regret is never having tried skydiving. I love high places! I always wanted to climb the ladder of a fire engine.

In contrast, Maiko-san was concerned that a smartphone might add to the social pressure she already felt was exacerbated by digital platforms. She talked about how on Facebook she felt compelled to 'like' posts by acquaintances in order to maintain smooth social relations. She used the popular term 'duty like' (*giri iine*) to describe this kind of action. When others liked her posts she hoped they were 'real likes' and not just 'duty likes'. The 'affective labour' (Hardt and Negri 2001) and 'emotion work' (Hochschild 1979) performed through visual digital practices, such as liking a Facebook post or sending an emoji or sticker, have, I argue, become central to the performance of 'care at a distance' (Pols 2012). On the one hand this form of care can become burdensome if deemed a requirement, as demonstrated by Maiko-san's negative feelings. On the other hand, visual digital media has affective potential that can serve to reduce feelings of burden in relationships of care, as will be demonstrated below.

Maiko-san was not against digital technologies per se. She accessed social media and her emails through her laptop, believing these were essential for communicating with her friends around the world, as well as with her students (she was a music teacher). She just felt no need to add to her digital repertoire with a new device. However, that all changed one day after a large typhoon struck Kyoto. I had returned from fieldwork to the UK at this time, but I suddenly received a Facebook message from Maiko-san saying 'Guess what: I've got a smartphone! I love it'.

I knew that I needed to follow up this development as remote ethnography, so I invited Maiko-san to collaborate on a video about her new smartphone. I interviewed Maiko-san via a Facebook video call, then wrote a video script based on the interview transcript that Maiko-san then recorded with her new phone. I then set this recording to video footage that I had previously recorded in Japan. The resulting video can be watched via the link in the image caption below (Fig. 5.3).

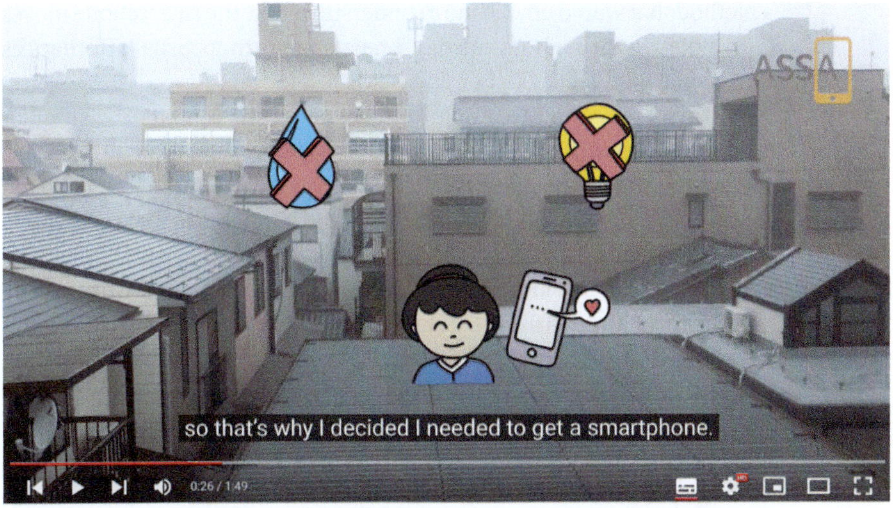

Figure 5.3 Watch the video, 'The smartphone is a lifeline', collaboratively made with Maiko-san here: https://youtu.be/z9AY2KWj4Rs?feature=shared.

This remote video-making methodology was an exercise in both data gathering and analysis. The initial interview provided the data, founded on an already existing research relationship with Maiko-san that had spanned my entire fieldwork in Japan. The analysis involved in selecting and arranging quotes into a narrative for a script was then made visible to Maiko-san. This provided an opportunity for her to reflect on what she had told me and my presentation of that information. This exercise demonstrated the possibilities that the smartphone opens up for remote research and collaboration with participants. I then expanded on these possibilities through the comic project I engaged in with Maiko-san, to which I now turn.

A graphic approach

The prevalence of illustrated media in contemporary life in Japan, including manga and now the daily exchange of billions of stickers on the messaging application LINE, means that as an anthropologist I needed to find a similarly visual means to explore the smartphone. Smartphones were used most by my interlocutors in private moments, away from my observation, so how might one 'follow the smartphone' (Marcus 1995)? Through conducting 25 smartphone interviews, I learned about the apps that people use and their feelings towards the device. However, to develop a method that would enable me to understand how the smartphone moves with a person throughout the day, I needed to turn to visual affordances of the smartphone itself – in this case the screenshot.

The production of the comic presented below was based on Maiko-san's digital screenshots, subsequently used as an elicitation device. This method was made possible because of shared digital practices between researcher and research participant. Multimodal research methods 'generate relations' and foster forms of collaboration that can make for a 'more public, more collaborative, more political' anthropology (Dattatreyan and Marrero-Guillamón, 2019, 221) – and I would argue also make empathetic engagement with research subjects more possible. At the core of my pursuit of graphic approaches is a desire to represent the digital practices of my interlocutors in a way that conveys how their digital repertoires develop in relation to things that matter in their lives.

To produce this comic, I first asked Maiko-san to keep a visual diary of her smartphone use for one day. She recorded her smartphone usage by taking screenshots each time she used her phone, resulting in 16 screenshots. We then discussed these together at length. I asked her questions, such as where she was when she took each screenshot, how she was feeling and what her motives were etc. This helped me to gain a comprehensive understanding of Maiko-san's usage for one day, building also on my knowledge of her daily habits gained through a year and a half of in-situ fieldwork.

What emerged was a clear insight into Maiko-san's use of the smartphone as part of an ecology of devices, or a 'network of personal technologies' (Fortunati and Taipale 2017), situating her within relationships of care. Her smartphone was used in parallel with her laptop, with Maiko-san taking advantage of the affordances of a bigger keyboard on her laptop for writing messages and of photos being stored on her phone for sending photos, as well as sending speech-to-text messages while she waited for her laptop to start up. In making the comic I was able to present a draft version back to Maiko-san who provided feedback and added details, which I then incorporated into the final draft. These additions included the thoughts she had while performing various tasks.

Maiko-san was keen to participate in this collaborative comic-making exercise. I had previously interviewed her several times about her technology use, sitting in her home and talking over tea, as well as via video calls as described above. I had also talked with her casually whenever the opportunity arose during my fieldwork about the role of communication technology in her life. Yet it was only by conducting this remote, comic-based method that I was able really to appreciate the significance of digital technologies in the private moments of her daily life. Presenting a visual narrative back to Maiko-san in a form that she could quickly understand and enthusiastically engage with enhanced the collaborative and participatory nature of this exercise. It also provided further insights regarding her thoughts and feelings which had not emerged in a prior interview with her about her smartphone.

Taking inspiration from the proliferation of visual communication among my interlocutors, producing a graphic narrative as a form of multimodal collaboration was a logical step; it enabled me to engage with them via a medium in which they were already literate. With the emergence of smartphones and other digital devices as ubiquitous across populations, for the first time anthropologists increasingly share access to the same means of representation and technological skills as their research participants (Favero and Theunissen 2018). Even when skill levels may not be comparable, the resulting difficulties or discrepancies can also be opportunities for insight, as occurred in the production of the comic below. Maiko-san did not initially know how to take a screenshot, and I was unfamiliar with her 'easy' smartphone Android device. Her immediate response was to search on Google for how to take a screenshot. This led to a discussion of how she often watches YouTube videos when she does not know how to do something.

As described in the video above (Fig. 5.3), during the typhoon that hit Kyoto in 2020 Maiko-san's home was left without electricity. She was therefore unable to access information via her laptop or TV. 'I felt blind,' she told me. She realised that she would feel more secure as an older person living alone if she had a smartphone with a data plan. She chose an 'easy smartphone' Android device, aimed at older adults, primarily because of the low cost. While acquiring the device was for practical reasons, Maiko-san soon discovered that she enjoyed using it. In particular she liked the speech-to-text capability which made messaging easier, given that the small keyboard and her eyesight had previously been a barrier to access. I was surprised at the complete reversal in her attitude. However it was only when engaging her in the comic exercise that I became fully aware of just how central a position the smartphone now had in her life.

The comic below (Figs. 5.4, 5.5, 5.6 and 5.7) shows how Maiko-san uses her laptop and smartphone simultaneously to send and reply to messages with text and stickers, send a photo, consume media and browse websites. Through this graphic form we also learn a range of other insights about Maiko-san's technology use, such as how she keeps her phone and laptop charging together overnight, that she does her

Figure 5.4 Maiko-san's day: page one. Image by the author.

Figure 5.5 Maiko-san's day: page two. Image by the author.

Figure 5.6 Maiko-san's day: page three. Image by the author.

Figure 5.7 Maiko-san's day: page four. Image by the author.

morning exercises while waiting for her laptop to start up and that she enjoys the warmth of the laptop on her lap as she sits in her armchair to scroll through the morning news. Her technology use predominantly takes place at home, except when taking a photo and reading messages.

We see how Maiko-san interchangeably sends messages via speech-to-text through her smartphone, replies to messages using the web interface of her messaging application and then sends a photo directly from her phone. Her physical proximity to her multiple devices, and the way in which she engages with them in the space of her home, is visualised; so is the audio notification when she receives a message. In this way, the graphic form enables a multisensory presentation of Maiko-san's morning routine with her devices in a manner that allows the viewer to dwell on each moment and absorb details, making connections for themselves between panels. As McCloud (1993) argues, it is in this imaginative work between the panels that the power of comics lies.

The comic shows Maiko-san embedded in a network of relationships of care. There is care between Sato-san and Maiko-san in their communication, care from commercial sources (the *bento* delivery service) and care that Maiko-san is showing to her unnamed friends by exchanging messages, songs and photos. We see how care is embodied in the way that Maiko-san and her teacher bow to each other when Maiko-san departs, which is then mirrored in the body language of the characters in the stickers.

There is also the self-care that Maiko-san is performing for her health – aided by checking her step counter, which motivates her to walk more the next day. The smartphone enables Maiko-san to be embedded in the practices of care which cannot be classified as 'online' or 'offline' but are instead shown here to be located physically and digitally simultaneously. Despite living alone, we see how Maiko-san remains close to others, with affective embodied gestures communicated through visual messaging.

In the process of making the comic, I discovered that Maiko-san often relies on her brother to drive her to medical appointments at the hospital. Their communication via LINE primarily revolves around organising these hospital trips, with Maiko-san using stickers, often of a bowing sheep character, to express her gratitude for his help, sometimes accompanied by the phrase '*itsumo suimasen*' ('thank you for your continued support') (Fig. 5.8). She finds stickers to be a convenient shorthand for efficiently conveying sentiments that would be cumbersome to type out in words. Maiko-san understands that her brother's busy schedule as a doctor warrants concise replies, often consisting of just one sticker to convey 'roger that' or 'understood'. This in turn allows for more efficient organisation of care and minimises any sense of burden that she may feel about relying on him.

More broadly among my research participants, I found that one must carefully judge the appropriateness of forms of communication according to the situation. Sometimes people felt it would be inappropriate to send a sticker; it might make the conversation appear too casual and even be thought disrespectful, depending on one's relationship to the recipient. Depending on the person one was messaging, people carefully crafted their messages to suit the social situation, 'polishing' their messages until they had acquired the correct form.

Being able to 'read the air' (*kuuki wo yomu*) accurately by paying attention to bodily expression and masking problematic feelings (in effect, paying attention to etiquette around one's public facade, or *honne*)[6] is an important part of Japanese social relations; it is indeed considered a sign of maturity (Cook 2013). My research revealed that a core concern behind hesitancy with digital communication among older adults was anxiety about 'reading the air' online. They perceived digital textual communication to be limited because of the impossibility of the non-verbal cues that they relied on in offline interpersonal communication. Elsewhere, I argue that it is this concern for exchanging messages that will

Figure 5.8 Conversations between Maiko-san and her brother featuring the use of a bowing sheep sticker. Image by the author.

not be misunderstood that has given rise to the explosion in digital visual communication evident today among Japanese older adults, especially women (Haapio-Kirk 2024). I propose that LINE stickers, such as the ones shared between Maiko-san and her brother, create 'atmosphere' which is essential for social relations more broadly.

Digital visual communication matters to Maiko-san, as do the other ways in which she relates to others through her devices. The graphic form allows a viewer to see the kinds of visual communication Maiko-san is engaged in an immersive format. Maiko-san was fairly representative of many participants of a similar age, who often combined the use of smartphones with personal computers (*pasokon*) and tablets. However, it was using the application LINE on her smartphone that afforded her particular visual communicative possibilities. I contend that emojis and stickers predominantly serve to maintain social connections through the 'emotion work' that they perform, rather than being used because of their inherent meaning (Wang and Haapio-Kirk 2021). This social function, which Malinowski terms 'phatic communication' (Malinowski 1946), can be understood as constituting the digital 'atmosphere' so crucial to offline communication. Elsewhere I have discussed how visual digital communication can mitigate some of the anxieties caused by lack of physical proximity in a rapidly depopulating Japan (Haapio-Kirk 2024).

Discussion

The concept of digital repertoires effectively highlights older adults' adaptability and creativity in engaging with technology. Contrary to stereotypes that often depict them as technology-averse, studies such as Kanayama's (2003) digital ethnography of senior virtual communities reveal that older adults have in fact been active adopters and innovators of digital technologies since the early days of the internet. Through participant observation conducted on a mailing list between 1999 and 2000, Kanayama found that the 120 members of this highly active group overcame limitations of text-based communication by experimenting with a variety of language forms, among them haiku and emoticons. The research demonstrated the propensity of older Japanese adults to find ways to overcome the lack of physical proximity that might otherwise afford emotional cues in digital communication.

Many older adults in my own research used their smartphones to maintain social connections, access information and engage in lifelong learning, shaping unique digital repertoires suited to their needs. The digital repertoires described in the case studies above showcase how offline and online forms of communication are often blurred, with both innovating each other. In this chapter I have endeavoured to show how the study of digital repertoires must adopt a holistic approach that captures the range of devices and practices that people simultaneously adopt, particularly in the communication of care.

For some research participants, the smartphone was their entry to the internet; they did not use computers at home. However, others adopted a blend of various technologies depending on the ease of use. Some preferred to compose emails or long messages on their computers, which they felt were easier to type on than smartphones. The smartphone was generally deemed to be the most convenient of all their devices because of its portable nature, as Sato-san explained:

> On a PC I have to lean in to look at the screen. But on this [the smartphone], it's so easy and I can make the screen and words bigger!

My research revealed that not only devices but also applications and services were chosen for their suitability for certain tasks. Emailing remained integral to how many people communicated, even with the rise of mobile messaging. People would use emails to communicate with friends and family, to share photos and to send 'important' messages. Messaging on the smartphone was convenient for informal regular messaging, but participants preferred to use email if they wanted to send something of significance. Sato-san explained:

> Email is more for longer sentences and things I would want to keep. Promises and things. Like, I ordered this and this is how many I ordered, things like that. So I can take a look later.

Yet seeing the 'read' notification on LINE was also reassuring for Sato-san, so if she sent an important email she would follow it up with a message to confirm that her communication had been received. Such findings regarding the screen ecologies (Miller et al. 2021) that emerge through personal choice and accessibility demonstrate the importance of a broad and inclusive approach to studying digital repertoires within everyday life.

The concept of digital repertoires offers a valuable approach to understanding the digital lives of older adults by framing their technology use as dynamic and adaptive. Rather than presenting older adults as passive consumers of technology, it invites an investigation into how they curate and refine their digital practices to fit their own lifestyles, preferences and social contexts. While the concept of digital repertoires effectively captures the strengths of older adults in adapting to, and innovating within, digital landscapes, it is also important not to downplay the significant barriers some older adults continue to face – such as varying levels of digital literacy and access to technology – as these can significantly influence their engagement. However, barriers to access are not age-bound. As I have shown above, factors of cost, privacy and emotional labour can lead to a hesitancy in digital adoption among people of all ages. The broad applicability of the digital repertoire approach must be complemented by analysis of the nuanced motivations behind digital adoption and interactions, which can vary widely across different cultural and social contexts.

Conclusion

This chapter has demonstrated, through case studies and reference to broader ethnographic findings, the ways in which people craft their digital repertoires in relation to practices of care. The concept of digital repertoire as a set of meaningful practices (Hänninen et al. 2023) is expanded to include spatial and affective dimensions, developed via a graphic methodology that seeks to reveal contextual data about private digital repertoires that may not otherwise be visible to researchers. In focusing upon the personal interests, skill sets and availability of technologies in my examination of the digital practices of older adults, I have shown specifically what older adults do with technologies. This in turn reveals how they manage challenges in their everyday lives, from natural disasters to affective communication.

This chapter demonstrates how the meaning of the term 'smartphone' is always contextual. To understand the smartphone in Japan, one must acknowledge the lineage from previous technologies such as *garakei.* It then becomes equally important to acknowledge the dominance of one specific app: LINE. In turn, while LINE can be used in a multitude of ways, a primary affordance is messages and the sharing of specific genres of visual media. The case studies discussed above reveal how such digital repertoires rely on an ecology of devices and apps used in tandem. The configuration of devices for Maiko-san, for example, is as bound up with their limitations as their affordances. Her laptop takes a long time to start up, explaining why her phone is the first device she uses while she waits. Her phone's keyboard is too small and difficult to type on, which is why she prefers using her laptop to send messages. Through the creation of the comic we saw how digital repertoires are developed in embodied and emplaced ways, focusing attention on the material dimension of technology use. The graphic methodology demonstrates how the smartphone affectively connects Maiko-san with others, enmeshing her in a web of care and support.

Through Maiko-san's screenshot diary I was able to build a detailed picture of one day in her life, complemented by many hours of participant observation as she attended *ikebana* lessons, spent time at home and went about her daily routines. The screenshots provided a window into the private moments of social interaction online that would have otherwise remained invisible. I was then able to illustrate these moments through a further visual experiment in the production of a graphic narrative. Presenting this narrative back to Maiko-san enabled even deeper opportunities for learning about how smartphone use is situated in affective embodied states and physical locations.

This graphic experimentation and the wider ethnography (Haapio-Kirk 2024) reveal not only how the adoption of the smartphone by older adults enables 'care at a distance' (Mol et al. 2010), but also underlines that in many cases care *requires* distance. This chapter has demonstrated how emerging digital repertoires among older adults become meaningful and central to their lives because of how they allow them to mitigate their sense of vulnerability and dependency on others and to manage affective labour, while simultaneously embedding them in relationships of care. I have tried to demonstrate a digital repertoire approach that highlights older adults' resourcefulness by acknowledging both their innovative practices and the structural challenges and cultural contexts that shape their digital experiences.

Acknowledgements

Sincere thanks to Sato-san and Maiko-san, along with everyone else who participated in my research. The research in this chapter was funded by the European Research Council (ERC) under the European Union's Horizon 2020 research and innovation programme (grant agreement No. 740472).

Notes

1 All participants are referred to using pseudonyms in order to protect their anonymity.
2 For example, in comparison with the UK or USA.

3 This finding is not limited to Japan, but is representative of a wider global trend among older adults. For example, in the US the past decade has seen a significant increase in the adoption of smartphones among the oldest age group cohorts, with the gap in usage between the oldest and youngest adults diminishing over time (Faverio 2022).

4 Mobile Society Research Institute 2021. 'Mobile Society White Paper web version. Chapter 6: Living conditions of seniors and use of ICT'. Accessed 3 February 2022. https://www.moba-ken.jp/whitepaper/21_chap6.html.

5 I interviewed Sato-san with my research assistant Lise Sasaki.

6 The concepts of *honne* ('true voice') and *tatemae* ('public facade') have a particular history in the post-war discourse on 'Japaneseness' (Lebra 2004). Both facets are important for engaging competently in interpersonal communication. As Goodman and Refsing (1992, 3) note, 'one of the first distinctions any anthropologist embarking on research in Japan learns to make is between *tatemae* and *honne*. *Tatemae* ... refers to an individual's explicitly stated principle, objective or promise; *honne* refers to what that individual is really going to do, or wants to do'. I recognise that these terms belong to essentialising discourse that Japanese people often have about themselves, associated with the wider '*Nihonjinron*' ('theories of Japaneseness') tradition of Japanese identity-making that emerged in the post-war period (Befu 2002). However, these words had clear meaning for my interlocutors, as became evident from the way they were reflected in their use of smartphones.

References

Akimoto, A. 2013. 'Looking at 2013's Japanese social-media scene', *The Japan Times*, 17 December 2013. Accessed 17 August 2021. https://www.japantimes.co.jp/life/2013/12/17/digital/looking-at-2013s-japanese-social-media-scene-3/.

Banks, M. and D. Zeitlyn. 2015. *Visual Methods in Social Research*. Second edition. London: SAGE Publications Ltd. https://doi.org/10.4135/9781473921702.

Befu, H. 2002. 'Symbols of nationalism and *nihonjinron*'. In *Ideology and Practice in Modern Japan*, edited by R. Goodman and K. Refsing, 40–60. Abingdon, Oxon; New York, NY: Routledge.

Byford, S. 2012. 'Docomo's Raku Raku Smartphone F12-D reinvents Android 4.0 for seniors (hands-on)', *The Verge*, 16 May 2012. Accessed 27 June 2021. https://www.theverge.com/2012/5/16/3023710/docomo-raku-raku-smartphone-f12-d-android-4-0-seniors.

Cook, E. E. 2013. 'Expectations of failure: Maturity and masculinity for freeters in contemporary Japan', *Social Science Japan Journal* 16(1): 29–43. https://doi.org/10.1093/ssjj/jys022.

Cox, R. A., A. Irving and C. Wright (eds). 2016. *Beyond Text? Critical practices and sensory anthropology*. Manchester: Manchester University Press.

Dattatreyan, E. and I. Marrero-Guillamón. 2019. 'Introduction: Multimodal anthropology and the politics of invention', *American Anthropologist* 121(1): 220–8. https://doi.org/10.1111/aman.13183.

Faverio, M. 2022. 'Share of those 65 and older who are tech users has grown in the past decade', *Pew Research Center*. 13 January 2022. https://www.pewresearch.org/short-reads/2022/01/13/share-of-those-65-and-older-who-are-tech-users-has-grown-in-the-past-decade/.

Favero, P. S. H. and E. Theunissen. 2018. 'With the smartphone as field assistant: Designing, making and testing *EthnoAlly*, a multimodal tool for conducting serendipitous ethnography in a multisensory world', *American Anthropologist* 120(1): 163–7. https://doi.org/10.1111/aman.12999.

Fortunati, L. and S. Taipale. 2017. 'Mobilities and the network of personal technologies: Refining the understanding of mobility structure', *Telematics and Informatics* 34(2): 560–8. http://dx.doi.org/10.1016/j.tele.2016.09.011.

Goodman, R. and K. Refsing (eds). 1992. *Ideology and Practice in Modern Japan*. Abingdon, Oxon; New York, NY: Routledge. https://doi.org/10.4324/9780203035283.

Goodwin, M. H. 2015. 'A care-full look at language, gender and embodied intimacy'. In *Shifting Visions: Gender and discourses*, edited by A. Jule, 27–48. Newcastle upon Tyne: Cambridge Scholars Publishing.

Haapio-Kirk, L. 2024. *Ageing with Smartphones in Japan: Care in a visual digital age*. London: UCL Press.

Hänninen, R., L. Pajula, V. Korpela and S. Taipale. 2023. 'Individual and shared digital repertoires – older adults managing digital services', *Information, Communication & Society* 26(3): 568–83. https://doi.org/10.1080/1369118X.2021.1954976.
Hardt, M. and A. Negri. 2001. *Empire*. Cambridge, MA: Harvard University Press.
Hasebrink, U. and H. Domeyer. 2012. 'Media repertoires as patterns of behaviour and as meaningful practices: A multimethod approach to media use in converging media environments', *Participations: Journal of Audience & Reception Studies* 9(2): 757–79.
Hendry, J. 1993. *Wrapping Culture: Politeness, presentation and power in Japan and other societies*. Oxford: Oxford University Press.
Hochschild, A. R. 1979. 'Emotion work, feeling rules, and social structure', *American Journal of Sociology* 85(3): 551–575. https://doi.org/10.1086/227049.
Ito, M. 2005. 'Mobile phones, Japanese youth, and the re-placement of social contact'. In *Mobile communications*, edited by S. Woolgar, R. Ling and P. E. Pedersen, 131–48. London: Springer.
Kanayama, T. 2003. 'Ethnographic research on the experience of Japanese elderly people online', *New Media & Society* 5(2): 267–88. https://doi.org/10.1177/1461444803005002007.
Kavedžija, I. 2018. 'Of manners and hedgehogs: Building closeness by maintaining distance' *The Australian Journal of Anthropology* 29(2): 146–57. https://doi.org/10.1111/taja.12274.
Lebra, T. S. 2004. *The Japanese Self in Cultural Logic*. Honolulu: University of Hawai'i Press. https://muse.jhu.edu/book/8167.
Malinowski, B. 1946 [1923]. 'The problem of meaning in primitive languages'. In *The Meaning of Meaning: A study of the influence of language upon thought and of the science of symbolism*, edited by C. K. Ogden and I. A. Richards, eighth edition, 296–336. New York: Harcourt, Brace & World, Inc.
Marcus, G. E. 1995. 'Ethnography in/of the world system: The emergence of multi-sited ethnography', *Annual Review of Anthropology* 24(1): 95–117. https://doi.org/10.1146/annurev.an.24.100195.000523.
McCloud, S. 1993. *Understanding Comics: The invisible art*. Princeton, WI: Kitchen Sink Press.
Miller, D., L. Abed Rabho, P. Awondo, M. de Vries, M. Duque, P. Garvey, L. Haapio-Kirk, C. Hawkins, A. Otaegui, S. Walton and X. Wang. 2021. *The Global Smartphone: Beyond a youth technology*. London: UCL Press. https://doi.org/10.14324/111.9781787359611.
Mol, A., I. Moser and J. Pols. 2010. 'Care: Putting practice into theory', *Care in Practice: On tinkering in clinics, homes and farms* 8: 7–27.
Murai, S. 2014. 'SoftBank unveils smartphone for seniors', *The Japan Times*, 28 October 2014. Accessed 10 December 2021. https://www.japantimes.co.jp/news/2014/10/28/business/corporate-business/softbank-unveils-smartphone-for-seniors/.
Okada, T. 2005. 'Youth culture and the shaping of Japanese mobile media: Personalization and the Keitai Internet as multimedia'. In *Personal, Portable, Pedestrian: Mobile phones in Japanese life*, edited by M. Itō, D. Okabe and M. Matsuda, 41–60. Cambridge, MA: MIT Press.
Pink, S. 2012. *Advances in Visual Methodology*. Thousand Oaks, CA: Sage Publications. https://doi.org/10.4135/9781446250921.
Pols, J. 2012. *Care at a Distance: On the closeness of technology*. Amsterdam: Amsterdam University Press. https://www.jstor.org/stable/j.ctt6wp5zw?pq-origsite=summon.
Schwarzenegger, C. 2020. 'Personal epistemologies of the media: Selective criticality, pragmatic trust and competence–confidence in navigating media repertoires in the digital age', *New Media & Society* 22(2): 361–77. https://doi.org/10.1177/1461444819856919.
Stoller, P. 1997. *Sensuous Scholarship*. Philadelphia, PA: University of Pennsylvania Press.
Thomas, J. M. and J. G. Correa. 2015. *Affective Labour: (Dis)assembling distance and difference*. Lanham, MD: Rowman & Littlefield International.
Wang, X. and L. Haapio-Kirk. 2021. 'Emotion work via digital visual communication: A comparative study between China and Japan', *Global Media and China* 6(3): 325–44. https://doi.org/10.1177/20594364211008044.

Part III

Mundane practices and visual methods

6

Ageing in digitalised Brazil: using the 'Brazilian way' to compensate for lower digital skills

Marília Duque and Emilene Zitkus

Introduction

During the Covid-19 pandemic, several services migrated online due to lockdowns or the limited physical access imposed to restrict virus transmission. This migration made evident that the ability to navigate the internet and use digital services are crucial skills for living in contemporary society. Furthermore, it made the digital divide in our societies clearly visible, especially in contexts of inequalities (van Deursen and Van Dijk 2011).

The digital divide in this study refers to the distinction between people who can access and use widely available technologies, the internet and the digital resources on the one hand, and those who are digitally excluded, unable to access, use or fully benefit from them on the other (Choudrie et al. 2013; Cunha 2022). While digitalisation appears to be a cost-efficient and citizen-oriented project (Hänninen et al. 2023), it can in fact result in social exclusion, challenging the autonomy of individuals with no or limited internet access and insufficient digital skills (Helsper 2016). Digital skills encompass a range of abilities required to use the internet and digital resources, such as mobile applications (apps) – increasingly necessary to access public services and online banking, or to undertake general communication, ensure online security and so on (Eshet-Alkalai 2004; van Laar et al. 2017; Gov.uk 2018; Lloyds 2021).

In this chapter, we present and discuss the findings of a 16-month ethnographic study. This research took place between 2018 and 2019 and involved a group of older adults living or using the services available in a neighbourhood of São Paulo, the most populated city in Brazil. The research aimed to investigate the strategies created by participants to compensate for their lack of digital skills when looking for health information, for example, or engaging with the digitalised institutional channels provided by public and private healthcare systems. The methodology section provides more information about the 38 participants of the study: their gender, age, levels of education, whether they were retired or still working and the type of healthcare service they used. This section also notes whether they were residents of the neighbourhood described above and if they used the services available there.

Given the extensive digitalisation of services in Brazil, however, older adults with limited or no digital skills have to overcome technological constraints to navigate the system. For example, participants rely on the favours provided by their social connections on WhatsApp, whose support shapes their digital repertoire and achievements online and offline. We approach these favours similarly to the 'Brazilian way' – a social, historical practice used in this case by Brazilians, but equally adopted by other countries around the globe, in order to overcome bureaucracies that seem impractical to comply with (Prado 2016). In relation to the concept of a digital repertoire, the 'Brazilian way' highlights the proactive role of Brazilian older adults in developing innovative approaches to utilising digital services for the purpose of enhancing everyday life.

In this way, this chapter focuses on the role played by friends with greater digital skills who somehow work as 'health resources'. Their part exceeds the roles of 'warm experts', who help older people to use the internet and craft their digital practices and repertoire, being a pivotal resource to technology adoption (Bakardjieva 2005; Duque 2022; Duque and Otaegui 2023; Hänninen 2023; Olsson and Viscovi 2018). Their healthcare background or personal influence in healthcare is how participants envisage health outcomes, from informal medical advice to preferential treatment. Although the ethnographic study was conducted before the Covid-19 pandemic, this chapter critically evaluates the use of digital services by older adults as part of the rapid transition to digital that was already occurring. It also reflects on some of the consequences of digitalisation for older adults[1] in Brazil, especially those possessing limited digital skills.

Background: Brazilian public encounters

Digital technology arises as a solution to optimise public encounters (Lindgren et al. 2019). It can make public services less bureaucratic, more cost- and time-efficient, open to citizen participation and accessible without discrimination (Freeman and Loo 2009; Lombardi and Vanolo 2015). The Brazilian strategy to digitalise public services has put pressure on public services, and the transition has occurred rapidly as a result. In 2022 the country was already among the GovTech leaders, according to the World Bank's GTMI. The index measures GovTech Maturity in four focus areas, including service delivery (World Bank 2022).

However, this was not the first attempt to use technology to improve public service delivery in the field site. 'Poupatempo' (meaning 'time-save') is an example of an earlier initiative. Created in 1997 by the government of São Paulo, 'Poupatempo' reshaped public encounters by speeding up the services offered by public authorities and utility companies. From applying for an identity card or renewing a driver's licence to managing bills for water supply, 'Poupatempo' encompassed more than 615 million requests for services between 1997 and February 2022 (Poupatempo 2022).

Prado and Trebilcock (2018) chose 'Poupatempo' to illustrate what they called an 'institutional bypass', referring to an institutional alternative for a dysfunctional institution. The 'institutional bypass' discharges functions performed by the dominant institution but does not replace it. Consequently, the institutional bypass is always relational to a dominant institution, implying that the actors can choose between them. That was the case with 'Poupatempo' as a new institution, PRODAM, was created to manage it while the dominant institutions providing the same services remained in place.

Institutional bypasses such as 'Poupatempo' can potentially change the nature of Brazilian public encounters. People no longer had to rely on informal strategies and personal connections to achieve the public services they need. In his book *O que faz o brasil, Brasil?* ('What makes Brazil, Brazil?') the anthropologist Roberto DaMatta (1986) argues that Brazilians developed a 'Brazilian way', a unique strategy to navigate through bureaucracies and comply with laws perceived as 'absurd' or 'impossible to obey'.

This 'Brazilian way' is a social and historical practice rooted in the informal and creative mechanisms through which ordinary people use personal favours to 'bypass' constraints and benefit from the system (Prado 2016). The practice exists across different social classes. Although often perceived as a form of corruption, the 'Brazilian way' does not necessarily result in material gains, but rather in the receipt of favours in various spheres of everyday life. Regarding the public one, institutional bypasses could reduce the use of 'Brazilian ways' in engaging with public services, as they make them accessible (and efficient) without discrimination. The transition of public services to digital services can do the same (Freeman and Loo 2009), once offered as a self-service (Lindgren et al. 2019).

However, there is a crucial distinction to be made between an institutional bypass (Prado and Trebilcock 2018) and the public services going online. Like the institutional bypass, digital services offer a valuable new feature: the same public service provided in person can be done online from anywhere, as long as the internet is available. However, the possibility of choosing *between them* can be compromised when public services are available exclusively online. In other words, digital services can add a valued feature to public encounters. Yet for them to work as institutional bypasses to in-person services, citizens should be able to choose between both options, which is not always the case.

Poupatempo can also illustrate how the choice implicit in the concept of an institutional bypass can be lost. In 2020, during the Covid-19 pandemic, the state of São Paulo launched the digital application 'Poupatempo Digital'[2]. However, among the services provided, 23 could be done exclusively through the app (SP News 2020). At the federal level, the Brazilian government also launched a series of apps and websites to support citizens during the Covid-19 crisis, with some of them making the digital route not an alternative but the *only* way to achieve public service. The application 'CAIXA Tem' was one example. The mobile app was mandatory for those applying to receive 'Ajuda Emergencial', which provides emergency financial support to individuals struggling during lockdown during the pandemic. Citizens who were in need of 'Ajuda Emergencial' potentially would not have broadband at home, or would have limited digital skills.

Furthermore, in 2020, the Brazilian government launched its digital strategy for the biennium 2020–2022. In this strategy it declared that 100 per cent of federal services should go online by the end of 2022 (Ministério da Economia 2020). In 2019, according to the government, 55 per cent of the 33,500 federal services were already available online. However, the 'non-digital' option, as Hänninen et al. (2023) noticed in Finland, is also becoming less visible or available. In other words, the digital solution cannot be seen as a digital bypass if citizens can no longer choose it as an alternative to services delivered in person.

Made exclusively or not, the digitalisation of public encounters implies that citizens would have to develop digital competencies to go online to achieve the desired outcomes. Under the umbrella of digital literacy, these competencies encompass technical, informational, communicational, cognitive, socio-emotional, strategic, transactional and problem-solving skills (Eshet-Alkalai 2004; van Deursen and van Dijk 2011; van Laar et al. 2017). Combined, they can shape offline and tangible outcomes from health decisions (Santana et al. 2011) to the way individuals make sense of and participate in the world they live (Cunha 2022; Junqueira et al. 2023).

The centrality of digital competencies can exacerbate inequalities as they are determined by socio-economic background, with education being a strong predictor of digital skills (van Dijk and van Deursen 2014). In addition, access to the internet should be seen as the first barrier that can lead to digital exclusion, followed by digital skills and outcomes achievements – the second and third levels of the digital divide (Helsper 2016). Although the digitalisation of public services is rooted in the promise of efficiency, it poses the question of *to whom* it is efficient. In the Brazilian context, and regarding the ageing population, we can start by addressing its potential for digital exclusion by highlighting that 16 per cent of Brazilians aged 60 and over are illiterate (IBGE 2023), and only 48 per cent have access to the internet (Cgi.br 2022).

The following section offers an overview of the digital divide in Brazil and how it affects the ageing population's interaction with the government. In it we present the qualitative data collected during the ethnography study, illustrating how digital literacy impacts the ways in which older adults in São Paulo access health information, healthcare services and medical guidance. The main argument of this chapter is that a lack of digital skills can compromise people's digital public encounters, driving them back to informality. As we will show, among their digital repertoires (Hänninen et al. 2023) WhatsApp enables them to keep using the 'Brazilian way' as an alternative to solve their everyday needs, even when a digital service is theoretically available.

Digital exclusion in the Brazilian ageing society

Brazil has the sixth largest older population in the world (United Nations Population Division 2022), with over 32 million people aged 60 and over. Less than half were connected to the internet before and after the pandemic (CGI.br 2022; PNAD 2019), according to the Brazilian

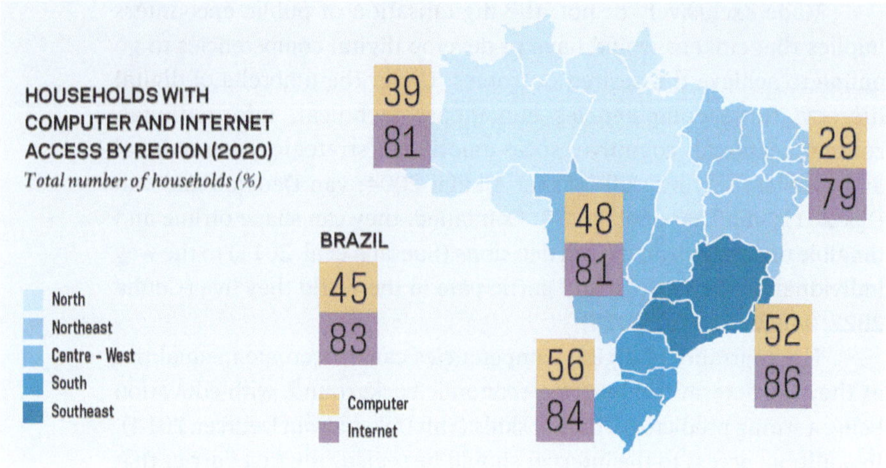

Figure 6.1 Internet access in Brazil: the percentage of households with a computer (in orange) and internet provision (in pink) by region in 2020. Image by Comitê Gestor da Internet no Brasil, 2021.

ICT Households Survey (PNAD). However, the share of Brazilians aged 60 and over using the internet has been growing fast, from 38 per cent in 2018 (PNAD 2018) to 45 per cent one year later; they are the age group with lower internet use (PNAD 2019). According to the most recent Brazilian ICT Households survey, 48 per cent of people aged 60 and over are connected, with most (80 per cent) accessing the internet through mobile phones (CGI.br 2022). The most frequently given reasons for being offline in Brazil relate to lack of connectivity, affordability, lack of the necessary skills and general attitude towards technology (CGI.br 2020).

Connectivity unavailability is related to infrastructural problems that reflect geographical and socio-economic inequalities. Internet provision or quality of availability varies significantly depending on the country's geographic regions, with less availability in rural or more deprived areas (Barbosa 2020; CGI.br 2021). Although infrastructural issues are changing quickly in Brazil, with connected households growing at an average rate of 4.3 percentage points per year – well above the world average of 2.7 per cent (CGI.br 2020) – the scenario remains problematic in several areas of the country. To illustrate regional differences, Figure 6.1 shows how the percentage of 'households with internet connection and computer at home' varies significantly among regions, with the north and northeast regions among the lower numbers.

The rates of internet use by older adults also reflects differences across the country. For example, only 28 per cent of the older adults in the north region (a socio-economically disadvantaged area, coloured light blue in Figure 6.1) had used the internet in the previous three months, according to a PNAD survey (2018). In contrast 47 per cent of older adults used the internet in the southeast (coloured dark blue in Figure 6.1) – the location of the empirical study presented here.

Another factor impacting access to the internet is the costs associated with it. According to the 16th edition of the Brazilian ICT Households survey (CGI.br 2021), the main barrier to internet access during the Covid-19 pandemic was the high cost of connection, a reason given by 68 per cent of respondents. According to the survey, in 2020 this comprised 'about nine million households without internet access in classes D and E, making up the bulk of the 12 million disconnected households' (CGI.br 2021, 220 §1). Although internet access costs tend to decrease with the expansion of internet provisions, the problem is the major one significantly affecting lower social classes – C, D and E (CGI.br 2021, 220).

According to the same survey, the other reasons for being offline are lack of digital skills (50 per cent), lack of need (49 per cent) or lack of interest (48 per cent) (CGI.br 2021). Interestingly, two of the three significant barriers – lack of need and lack of interest – call for a better understanding once digital skills became crucial to reach services and benefits during the pandemic. They have become almost mandatory in a 'compulsory computing society' such as the one we live in (Allmann and Blank 2021). In this scenario people are digitally connected but digitally excluded, as they lack the skills to navigate the digital world confidently and safely (Centre for Ageing Better 2021; Lloyds 2021).

Consequently, digital exclusion can be another driver for social exclusion, with citizens' autonomy being increasingly mediated by technology (Duque and Oliveira 2022). The Brazilian ICT Households (CGI.br 2021) survey illustrates how digital inequalities in Brazil can profoundly impact how older adults access and benefit from services. Only 53 per cent of older adults (that is, those aged 60 and over) interacted with a governmental institution online during 2020, while only 29 per cent used an online public service during the same period (CGI.br 2021, 239). Regarding healthcare, only 49 per cent searched the internet for health information or assistance.

Lack of digital skills and the impact on healthcare

Telemedicine 2.0 – including the ability to conduct medical appointments remotely – was not widely regulated before the Covid-19 pandemic. Teleorientation, teleconsultation and telemonitoring were then approved on an emergency basis in March 2020 (Zitkus and Libânio 2021). Both in the public and private spheres, a series of applications was developed or optimised to provide health resources, including apps such as the CoronavirusSUS and the ConecteSUS, both linked to the universal public health system in Brazil (Sistema Único de Saúde – SUS). Even considering that older adults were the age group most at risk from Covid-19 (Camarano 2020), these initiatives were not designed to respond to the needs of this specific age group. Therefore, as Zitkus and Libânio (2021) argue, older people may have both limited access to these resources and less ability to use them, as they might not have developed adequate digital skills to navigate the interfaces.

Helbostad et al. (2017) endorse the potential of mobile health (mHealth) to combine smartphones, sensors and applications, and to structure a new health paradigm for older people. These 'promising' technologies could encourage behaviour change and empower older people to extend their autonomy based on their ability to monitor, intervene and take charge of their health. However, they consider the evidence of its efficiency 'sparse'. Focusing on the older generation, they stress that these applications do not meet older people's needs and preferences, especially regarding functionalities, choice and usability. In addition to decreased sensory perception and poorly designed interface, Cajita et al. (2018) also consider the barriers to older people's engagement: lack of knowledge regarding how to use mHealth, lack of need for technology, the cost of technology for those on a limited/fixed income. The barriers confirm the reasons older adults in Brazil presented for staying offline, as discussed earlier in this chapter, as well as the need to bear in mind attitudes, preferences and skins when developing digital public encounters.

This chapter contributes to this discussion by providing a qualitative approach to how older adults in São Paulo engage with digital healthcare services and medical guidance, revealing the alternative routes they found to use their digital repertoires for health purposes.

Methodology: an ethnographic approach to mobile technology impacts on health and ageing

This ethnographic study was part of a collaboration with 'ASSA: Anthropology of Smartphones and Smart Ageing' at University College London (Duque 2021; 2022) which combined participant observation and in-depth interviews. Fieldwork was conducted from February 2018 to June 2019 in a middle-class district in São Paulo, an area with the second-highest rates of income, life expectancy and education among the 96 districts of the city (Gonçalves and Maeda 2017). In 2019 the proportion of residents aged 60 and over in this district was 25.1 per cent. The number of people aged 60 and over for every 100 people younger than 15 years old was 195.2 – both higher figures than the average for São Paulo (Secretaria Municipal de Direitos Humanos & Cidadania and Coordenadoria de Políticas para Pessoa Idosa 2019).

Participants should live in the area or attend local activities under the Active Ageing agenda (WHO 2002), adopted as a public policy in São Paulo and Brazil (Dias and Pais-Ribeiro 2018). During fieldwork, Duque joined Pilates and yoga classes, a meditation group and a group focused on senior entrepreneurship; she participated in community events and volunteered as a teacher in WhatsApp workshops at a Catholic church for three terms (lasting from four to five months each). The purpose of her presence and the aim of the research were disclosed to the groups, including those on WhatsApp.

The person in charge of the activities usually creates a WhatsApp group to support members. Duque was also accepted in those spaces, which allowed participant observation to be expanded to online encounters. During the fieldwork 10 WhatsApp groups were observed (Duque managed three to support her WhatsApp students). The smaller one had 12 members, the larger one almost 200. As participants combine multiple activities during the week, they are included in many WhatsApp groups that favour the emergence of a constantly expanding network.

Some of the participants of the activities observed during fieldwork were invited for an in-depth interview. They also referred relatives and peers to the study, according to the eligibility criteria. Interviews addressed their digital technology adoption and the impacts of smartphones on ageing and health. Participants using smartphones were also invited to show the apps installed and comment on their usage. An informed consent form was applied (Ethical approval via Certificate of Presentation for Ethical Consideration [CAAE] #90142318.2.0000.5511). A total of 38 interviews were translated and pseudonyms replaced participants' names.

As Miller et al. (2021) observed in their global comparative study on smartphones and ageing, what older people say and do about technology can be contradictory. The participant observation and in-depth interviews combined in this ethnography aimed to approach and compare discourses, practices, attitudes and behaviours towards technology. An attitude towards an object is not necessarily predictive of behaviour, of course. Behaviour results from the influence of demographic profile, personal characteristics, beliefs about the object, context necessary for action and situational variables (Hartwick and Barki 1994). This set of factors influences and shapes the way in which participants engage or not with digital solutions and their digital repertoire.

Participants' backgrounds and their digital repertoires

Among the 38 participants interviewed there were 22 women and 16 men. The age range was from 50 to 76. Most (30) had received higher education; 28 participants had children, 27 had health insurance and 28 were retired. Of the 10 who were not retired, seven were still working and three were unemployed.

All participants owned a smartphone and 17 had second-hand devices, presenting issues with battery and memory that sometimes compromised the functionality of apps or the ability to download them at all. All except one of the participants (37) had a data package plan. The exception was a retired woman who did not have a data package plan at the time of the interview. However, she bought one after attending a WhatsApp workshop and became more confident about using her smartphone.

Out of 38 interviewed participants, 27 had private health insurance. Of these, 13 had downgraded their health insurance policy after retirement. Prevent Senior, more affordable healthcare insurance targeted at older adults, was the new healthcare provider chosen by 100 per cent of those who opted to downgrade their private insurance plans. Nine relied exclusively on the Public Health System, known as the Sistema Único de Saude (SUS) and used by 75.3 per cent of Brazilians aged 60 or over (Penido 2018).

The digital repertoires of older adults in this study mainly evolved from their use of messaging apps. In our research we found that WhatsApp was already installed on all the participants' devices. The motivation for older people to adopt messaging apps such as WhatsApp is a direct consequence of its use in communication between nuclear, extended and transnational families (Ahlin 2018; Duque 2022; Gonzalez and Katz 2016; Nedelcu 2017; Plaza and Plaza 2019). WhatsApp is Brazil's most popular messaging app, which mediates everyday communication with family and friends (Mobile Time and Opinion Box 2021). It is thus the first app that participants were interested in adopting. Below we describe how the digital repertories of the participants were shaped and consider the role played by personal connections in supporting or compensating for the digital repertoires they use to achieve health outputs from public and private healthcare systems.

The role of 'warm' and 'cold experts' in the development of digital skills

We expected to find that close family members would act as 'warm experts', supporting older parents in their technological needs (Hänninen et al. 2023; Olsson and Viscovi 2018). However, this ethnography highlighted that most of the participants with children (28) complained they could not count on them. They argued that their children had neither the time nor the patience to help them with their digital needs, often perceived by the younger generation as a burden (Duque and Otaegui 2023). This explains why participants seek support from professionals with whom they have no personal connection, known as 'cold experts' (Hänninen et al. 2023). The demand for cold experts' advice was observed during fieldwork, with WhatsApp courses proving the most popular of the activities offered by the Active Ageing agenda in São Paulo.

The role of professional background and retirement age

In addition to social support, the age of retirement strongly influenced participants' digital competences and practices that are reflected in their digital repertoires. There was a clear gap between those who had been retired for 20, 10, or just a few years with the last being more digitally competent than those who had retired earlier. This in turn suggests that working environments now facilitate tech adoption (Duque 2022). On the other hand, as highlighted by Rivinen (2021), the work performed by older adults during their professional lives does not necessarily need to engage with technology, a fact also reflected in some cases in the ethnography study.

The reliable sources of online information

Participants tend to judge the reliability of the content they receive on WhatsApp based on the person from whom they received the information in the first place. If they trust the person, they are more likely to have confidence in the content (Duque and Peres-Neto 2022), an everyday life practice incorporated into the digital sphere. According to a DataSenado survey (2019), 24 per cent of Brazilians do the same. The trust attributed to personal connections compensates for participants' lack of media literacy, here meaning the ability to create, consume and engage critically with digital content.

Despite their educational background, personal connections also became a crucial resource for participants seeking access to health information and medical guidance they trust. This was particularly the case for those who struggled in navigating new 'geographies of care', where digital technology mediates healthcare and redistributes responsibilities over one's health (Oudshoorn 2011).

Managing health-related apps

During the interviews, three healthcare providers were mentioned by participants: SUS (Brazilian public healthcare system), Prevent Senior (private health insurance) and Dr Consulta (pay-per-use clinic). All of them provide apps for their users. However, participants' engagement with health-related apps during fieldwork was limited, despite their educational and socio-economic backgrounds. SUS's app 'Agenda Fácil', for example, was created in 2017 to make booking appointments and tests in the public healthcare system more effective and efficient (Prodam 2018). Yet of the 11 interviewed participants who used the SUS, only two knew that the 'Agenda Fácil' app existed.

The SUS app

One participant who knew about the SUS app was Sandra, a woman aged 69 who gave up using the Agenda Fácil app in the set-up phase. She was supposed to go to her nearest general practice unit – Unidade Básica de Saúde (UBS) – and ask for a code to validate her account in the app. She had problems getting her code, so she abandoned the app.

The other participant was Renato, a man aged 63. He could set up the app and considered the Agenda Fácil 'easy to use'. However, he also complained about the system, arguing that if the professionals at his UBS fail to update the system with the referrals and clinical tests needed, the app will not work. Should that happen, Renato would have to go in person to schedule an appointment. That inconvenience made him think of 'empowerment' as an institutional discourse that, in practice, does not work as advertised. This institutional distrust reinforces one of the outcomes highlighted in another study on older adults' digital engagement in the Brazilian context (Junqueira et al. 2023).

One of the features offered by Agenda Fácil is a digital alternative to the printed SUS ID, which allows people to use the services. As Sandra mentioned earlier, she could not use the app, so she came up with a creative strategy to compensate for it. She took a picture of her printed SUS ID and saved it in Google Drive, enabling her to present the image whenever needed. That is a behaviour shared by participants. They tend to return to the apps they know when they struggle to use new ones.

The Prevent Senior and Dr Consulta apps

The Prevent Senior app also provides a digital ID for users. This was the main feature and sometimes the only one used by the 13 participants who relied on this private health insurance. As the app is perceived as 'hard to use', they prefer to access the company website or use the telephone to schedule appointments and clinical tests (even if both functionalities are also available on the app). The same situation occurs with the low-cost pay-per-use clinic Dr Consulta.

The challenges the apps can represent

The shift to other ways to access the service provided by an application illustrates how a digital service can fail to meet older adults' digital skills and needs. Instead of benefiting from quicker access to health resources, participants may search for a bigger screen or even abandon the internet to use the telephone instead. This behaviour is also relevant if we consider the digital divide in the Brazilian context. While 89 per cent of the high-income class combine computers and smartphones to access the internet, only 10 per cent of the population from lower-income classes have multiple devices. In the case of users aged 60 and over, the majority of 64 per cent access the internet exclusively through their smartphones (CGI.br 2021). In these instances, the telephone can be seen as the only alternative they have when they struggle to use the app developed by a healthcare provider.

In addition, even the more digitally skilled participants, for example Sandra and Renato discussed above, may struggle to benefit from these apps. Rocha and Padovani (2016) argue that developers often ignore motor and cognitive changes naturally associated with ageing, resulting in poor usability of the digital service. The authors point to flaws in the design of interfaces that do not contribute to the context of the older user, in addition to failing to provide feedback on the actions performed, which may result in older people becoming disoriented during the interaction with smartphones. Fernanda, a highly educated woman aged 63, illustrates how poor interfaces compromise user experiences. She uses mobile banking apps, e-commerce apps and mobility apps. However, when she has to schedule an appointment with a doctor she prefers to use the Prevent Senior website on her desktop, not her smartphone.

Apart from the apps made available by health providers from the public and private sectors, participants could also engage with m-Health apps to exercise, control their diets or monitor their health status. However, only 13 out of 38 participants have apps such as these installed on their smartphones – which does not in itself mean they were using them. Only three participants engaged with these apps daily, and they did it in conjunction with smartwatches. Furthermore, only one participant, a woman aged 70 who holds a university degree, learned from and followed the app feedback.

Among the m-Health apps installed and never used are those pre-installed with the device, including those downloaded by previous owners of second-hand devices. When participants themselves downloaded the apps, a lack of trust in the metrics and feedback provided by the health applications contributed to them losing interest in or abandoning the health goal associated with it. In addition, three of those who had installed m-Health apps did it 'for future use'. All of these were diet-related apps.

We might argue that participants are not particularly interested in self-tracking apps. Participants often say they are 'lazy' or 'don't have time' to learn any new apps. The excuse is used when they do not have the skills to download and use a new app. This declared lack of interest covers their lack of ability or support, mainly when they cannot rely on their children as 'warm' experts. Asking for help could be embarrassing for participants and so deterred them from trying new apps. In that sense a 'lack of interest' could work as a self-preservation strategy (Duque and Otaegui 2023). These findings can contribute to understanding why 'lack of interest' was the main reason presented by Brazilians aged 60 and over not to connect to the internet, as highlighted above in the discussion of the 'digital divide'.

The following sections show how friends connected on WhatsApp emerge as the most valuable resource when participants try to access health information and medical guidance, and explore the impact of this on their digital repertoire (Hänninen et al. 2023). As we will demonstrate, friends do not play the role of warm experts. They are not committed to supporting their peers in developing their digital skills to expand their usage of online health services. However, their network on WhatsApp does offer an alternative support, providing tangible outcomes for those with low levels of digital literacy.

WhatsApp as an alternative route to access health information and healthcare

Participants often argue that when they have a health issue they are more likely to abandon searches on Google as they do not know how to evaluate the reliability of the information available online. In addition, they maintain deference to medical authority despite their often high levels of education. As Maria, a university graduate of 70, observed, 'there is a lot of misleading information online. I prefer to follow my doctor's recommendation'. Eduardo, a 59-year-old man who completed primary school only, knows the feeling:

> The internet can be a little bit scary, right? It sends you to a wild place with so much information you get lost. For example, I once sought information about something my wife was feeling. God, it was too much information. I told myself 'No – I will get out of here'. I will ask the doctor directly and he will give me the right information. I don't like it [the internet]. I did it once. It scared me more than it helped me.

Maria has private health insurance, which she complements with a dermatologist. As she pays for every consultation, she can reach this doctor relatively easily on WhatsApp. Eduardo relies on the public healthcare system, however, and has no access to his doctor's WhatsApp. He compensates for that by searching for reliable health information among his social connections on WhatsApp. When he and his wife were diagnosed with fatty liver disease, for example, Eduardo sought information about healthy diets. A friend of his provides him with daily information he can understand and trust, as this friend works in the healthcare area and is perceived as a 'qualified connection'. It became apparent from the fieldwork that such networks are maintained even when participants have access to their doctors' contact on WhatsApp. They tend to keep doctors' connections for 'real emergencies'. Furthermore, participants used their 'qualified connection' on WhatsApp for everyday needs to avoid being seen as a burden by doctors.

That was the case with Lourdes, a woman aged 64 who also held a university degree. She came across the information that older people should have the herpes zoster vaccine, but was not sure whether that applied to her or not. She already had an appointment with her doctor in two weeks, so did not want to bother him before that. In Lourdes's case she has her doctor's WhatsApp contact – a privilege acquired through being a patient of his father for many years. She had to stop seeing the father when her insurance plan coverage changed, but he referred her to his son, giving her his WhatsApp contact details as a 'bonus' based on their past relationship.

However, as already noted, participants keep such connections for emergencies only. Normally they prefer to seek informal but professional guidance from people they know personally. Lourdes thus contacted a friend on WhatsApp and asked her to ask her son whether she should have the vaccine. It did not take her long to get an answer. Lourdes explained that her friend's son was not just a doctor, but an infectologist working in one of São Paulo's most respected hospitals.

On the one hand, WhatsApp clearly facilitates communication among peers. On the other hand, it expands social connections, helping older adults to reach the qualified link associated with a specific field of interest or need. The cases described above show how skilled, health-related connections can compensate for participants' own lack of digital health literacy (Kickbusch et al. 2013). The messaging app mediates support from close connections, providing informal knowledge (based on the favour of a friend) and professional expertise (as this friend has expertise in healthcare). In addition, personal connections on WhatsApp are used to receive preferential treatment in public and private healthcare systems. The examples that follow demonstrate how WhatsApp can, in other words, facilitate the 'Brazilian way'.

João, a man aged 64 suffering from a rare autoimmune disease, illustrates the earlier case. As his own doctors were unable to provide a diagnosis, his wife sent a message to a friend from her church who was also a doctor. This friend asked the couple to go to the public hospital where she works, instructing them to message her as soon as they arrived. She met them at reception, let them in and introduced them to her colleague, a specialist in the area. João was immediately taken on as a patient and could start his treatment without queuing, something that would be impossible without leveraging such personal connections.

Knowing an 'insider' can also result in privileges within the private healthcare system. This scenario is illustrated by the case of Bia, a woman aged 55 whose father had died two years earlier. When her father was hospitalised, Bia's main concern was not his death but providing him with the dignity and comfort he deserved at the end of his life. She could not bear the fact that her father might die in a cubicle without windows. Bia therefore contacted a friend whose husband worked at a senior level for her father's health insurance company. In the company's hierarchy, she explains, 'He is on a par with the Pope'. Her father was then transferred to a bigger, brighter room. Bia also benefited from exceptional support from the palliative care team, constantly updated by the doctor in charge through WhatsApp.

Conclusion: the 'Brazilian way' through digital platforms

Institutional bypasses create alternative routes to bureaucratic institutions, making their services more efficient (Prado and Trebilcock 2018). Like 'Poupatempo' in Brazil, these routes can provide standardised treatment to all citizens. Digital public services emerge with the same, less bureaucratic and accessible (without discrimination) potential. In the Brazilian context this could mitigate the 'Brazilian way', a historical cultural practice that relied upon personal connections and influence to navigate the bureaucracies of the system.

However, digital services depend on individuals' access to the internet, digital competence and repertoire to become inclusive and convenient and to enhance daily life. If these factors are lacking, such services can create a new space for inequality and exclusion, especially for older people. This ethnographic study showed that, even when older adults have high levels of education, live or use services in a privileged neighbourhood and are connected online, their limited digital skills, outdated devices and lack of support from 'warm' experts impact on their competencies to achieve health outcomes. As mentioned in Chapter 1, individuals use the knowledge they have to navigate new contexts and challenges, incorporating non-digital aspects of everyday life into their digital practices, forming their digital repertoire.

On the one hand, we could observe restricted use of the health-related apps offered by public and private healthcare systems (Agenda Fácil, Prevent Senior and Dr Consulta). On the other, WhatsApp, a messaging app, played a crucial role in enabling a network of reciprocity-based favours where participants could search for qualified friends to access health information and medical guidance they could trust. We argue that these qualified connections compensate for participants' lack of digital skills to search for health information online. In addition to this, WhatsApp facilitated contact with influential connections, enabling the 'Brazilian way' to achieve preferential healthcare services in informal ways. In one way or another, we should recognise personal connections and qualified friends in healthcare as part of participants' digital repertoires used to achieve health outcomes.

By showing these ethnographic examples, it is crucial to consider that most of the participants have high levels of education and income and live in a privileged, age-friendly neighbourhood of São Paulo where the fieldwork was located. The outcomes could be completely different in other areas of the country or even in the same city, considering the inequalities among the 96 districts of São Paulo (Rede Nossa São Paulo 2020).

Bearing in mind the digital inequalities in the country, we conclude that the digitalisation of services in Brazil must be carefully addressed. Without such an approach digitalisation can lead to exclusion or drive older adults back to informality.

It is therefore important in developing new digital services to consider users' needs and expectations, as well as comfortable ways of using technology, which collectively constitute their individual digital repertoires. Furthermore, additional research could investigate more accessible methods to promote the digital inclusion of older adults, by supporting the digital skills and thus the digital repertoires that contribute to wellbeing in everyday life.

Acknowledgements

Duque's research was funded by CAPES, PDSE Program, Grant No. 88881.362032/2019-01. Ethical Approval: CAAE 90142318.2.0000.5511. Zitkus's work was supported by the Royal Society under Grant No. IES\R3\213004; Information Design Association (IDA); FAPESP – the Sao Paulo State Research Foundation under Grant 2020/05187-5.

Notes

1 We refer to older adults as people aged 50 and over for whom the internet has not been part of their formal education or placement work training. According to Hill et al. (2015), when discussing the use of digital resources, older adults are individuals aged over 50 years old if they did not use the internet during their formative adult years. The age of 50+ as a threshold has been used as a reference in other studies in the literature (Hill et al. 2015; Pan and Jordan-Marsh 2010; Peral et al. 2014), while others consider older adults those of retirement age, such as 60 (Heart and Kalderon 2013; Nimrod 2018; Selwyn 2004) or 65 years old (Barnard et al. 2013; Gell et al. 2015; Niehaves and Plattfaut 2014; Olson et al. 2010). They form the majority of the disconnected and less digital skilled people in the country (CGI.br 2022).

2 https://www.poupatempo.sp.gov.br/.

References

Ahlin, T. 2018. 'Only near is dear? Doing elderly care with everyday ICTs in Indian transnational families', *Medical Anthropology Quarterly* 32(1): 85–102. https://doi.org/10.1111/maq.12404.

Alkalai, Y. 2004. 'Digital literacy: A conceptual framework for survival skills in the digital era', *Journal of Educational Multimedia and Hypermedia* 13(1): 93–106.

Allmann, K. and G. Blank. 2021. 'Rethinking digital skills in the era of compulsory computing: Methods, measurement, policy and theory', *Information, Communication and Society* 24(5): 633–48. https://doi.org/10.1080/1369118X.2021.1874475.

Bakardjieva, M. 2005. *Internet Society: The internet in everyday life*. London: Sage Publications. https://doi.org/10.4135/9781446215616.

Barbosa, B. 2020. 'Brazil in the pandemic: Lack of quality connection and increasing inequalities'. In *2020 ICT Households: Survey on the use of information and communication Technologies in Brazilian households*, 248–57. CGI.br. Accessed 10 January 2025. https://cetic.br/media/docs/publicacoes/2/20211124201233/tic_domicilios_2020_livro_eletronico.pdf.

Barnard, Y., M. D. Bradley, F. Hodgson and A. D. Lloyd. 2013. 'Learning to use new technologies by older adults: Perceived difficulties, experimentation behaviour and usability', *Computers in Human Behavior* 29(4): 1715–24.

Cajita, M. I., N. A. Hodgson, K. W. Lam, S. Yoo, S. and H.-R. Han. 2018. 'Facilitators of and barriers to mHealth adoption in older adults with heart failure', *CIN: Computers, Informatics, Nursing* 36(8): 376–82. https://doi.org/10.1097/CIN.0000000000000442.

Camarano, A. A. 2020. 'Os dependentes da renda dos idosos e o coronavírus: Órfãos ou novos pobres?' ['The income dependants of the elderly and the coronavirus: orphans or the new poor?'], *Ciência & Saúde Coletiva* 25(2): 4169–76. https://doi.org/10.1590/1413-812320202510.2.30042020.

Centre for Ageing Better. 2021. *COVID-19 and the Digital Divide: Supporting digital inclusion and skills during the pandemic and beyond*. Report, 6 July 2021. Accessed 10 January 2025. https://ageing-better.org.uk/resources/covid-19-and-digital-divide-supporting-digital-inclusion-and-skills.

Choudrie, J., G. Ghinea and V. N. Songonuga. 2013. 'Silver surfers, e-government and the digital divide: An exploratory study of UK local authority websites and older citizens', *Interacting with Computers* 25(6): 417–42. https://doi.org/10.1093/iwc/iws020.

Comitê Gestor da Internet no Brasil, CGI.br (ed.). 2020. *ICT Households 2019: Survey on the use of information and communication technologies in Brazilian households*, first edition. Cetic.br & NIC.br. Accessed 10 January 2025. https://cetic.br/media/docs/publicacoes/2/20201123121817/tic_dom_2019_livro_eletronico.pdf.

Comitê Gestor da Internet no Brasil, CGI.br (ed.). 2021. *ICT Households 2020: Survey on the use of information and communication technologies in Brazilian households, COVID-19 edition: adapted methodology*. Cetic.br & NIC.br. Accessed 10 January 2025. https://cetic.br/media/docs/publicacoes/2/20211124200326/tic_educacao_2020_livro_eletronico.pdf.

Comitê Gestor da Internet no Brasil, CGI.br (ed.). 2022. *ICT Electronic Government: Survey on the use of information and communication technologies in the Brazilian public sector*. NIC. br. and Cetic. br. Accessed 10 January 2025. https://cetic.br/media/docs/publicacoes/2/20220725170710/tic_governo_eletronico_2021_livro_eletronico.pdf.

Cunha, I. M. R. F. 2022. 'Literacias para a cidadania global' ['Global citizenship literacies'], *Comunicação & Educação* 27(2): 183–98. https://doi.org/10.11606/issn.2316-9125.v27i2p183-198.

DaMatta, R. 1986. *O que faz o brasil, Brasil?* ['What makes Brazil, Brazil?']. Rio de Janeiro: Rocco.

DataSenado. 2019. *Redes Sociais, Notícias Falsas e Privacidade de Dados na Internet* [*Social Networks, Fake News and Data Privacy on the Internet*]. Report. Câmara dos Deputados & Senado Federal. Accessed 10 January 2025. https://www12.senado.leg.br/institucional/datasenado/arquivos/mais-de-80-dos-brasileiros-acreditam-que-redes-sociais-influenciam-muito-a-opiniao-das-pessoas.

Dias, E. N. and J. L. Pais-Ribeiro. 2018. 'Public policy evolution related to older people in Brazil' (original title: 'Evolução das políticas públicas á pessoa idosa no Brasil'), *Enfermagem Brasil* 17(4): 413–20. https://doi.org/10.33233/eb.v17i4.860.

Duque, M. 2022. *Ageing with Smartphones in Urban Brazil: A work in progress*. London: UCL Press.

Duque, M. 2021. 'Homo resiliens: Moralidades e resistências da velhice mediada por smartphones em São Paulo' ['Moralities and resistances of old age mediated by smartphones in São Paulo']. PhD thesis. Escola Superior de Propaganda e Marketing. Sao Paulo: Instituto Brasileiro de Informação em Ciência e Tecnologia. Accessed 10 January 2025. http://tede2.espm.br/handle/tede/566.

Duque, M. and A. L. de Oliveira. 2022. 'The accidental age-friendly city: Public expectation and subjective experience in São Paulo', *Revista Lusófona de Estudos Culturais* 9(1): 67–86. https://doi.org/10.21814/rlec.3649.

Duque, M. and L. Peres-Neto. 2022. 'Can older people stop sharing? An ethnographic study on fake news and active ageing in Brazil', *Online Media and Global Communication* 1(3): 580–99. https://doi.org/10.1515/omgc-2022-0034.

Duque, M. and A. Otaegui. 2023. 'Digital dependency as a burden: Impact of active aging for technology adoption in Brazil and Chile', *Anthropology & Aging* 44(2): 28–44. https://doi.org/10.5195/aa.2023.422.

Freeman, R. J. and P. Loo. 2009. 'Web 2.0 and e-government at the municipal level'. In *2009 World Congress on Privacy, Security, Trust and the Management of e-Business*, 70–8. St. John's, NL, Canada: IEEE. https://doi.org/10.1109/CONGRESS.2009.26.

Gell, N. M., D. E. Rosenberg, G. Demiris, A. Z. LaCroix and K. V. Patel. 2015. 'Patterns of technology use among older adults with and without disabilities', *The Gerontologist* 55(3): 412–21. https://doi.org/10.1093/geront/gnt166.

Gonçalves, A. de F. and M. T. Maeda. 2017. 'IDH E a dinâmica Intraurbana na Cidade de São Paulo' ['HDI and intra-urban dynamics in the city of São Paulo']. In *Territórios em números: Insumos para políticas públicas a partir da análise do IDHM e do IVS de UDHs e regiões metropolitanas brasileiras [Territories in numbers: Inputs for public policies based on the analysis of the Municipal Human Development Index (MHDI) and the Social Vulnerability Index (SVI) of Human Development Units (HDUs) and Brazilian metropolitan regions*, edited by B. O. Marguti, M. A. Costa and C. V. da S. Pinto, vol. 2, 171–91. Brasilia: IPEA. https://portalantigo.ipea.gov.br/portal/images/stories/PDFs/livros/livros/170828_livro_territorios_numeros_insumos_politicas_publicas_2_cap06.pdf.

Gonzalez, C. and V. S. Katz. 2016. 'Transnational family communication as a driver of technology adoption', *International Journal of Communication* 10: 2683–2703.

Gov.uk. 2018. *Essential Digital Skills Framework*. Accessed 10 January 2025. https://assets.publishing.service.gov.uk/government/uploads/system/uploads/attachment_data/file/738922/Essential_digital_skills_framework.pdf.

Hänninen, R., L. Pajula, V. Korpela and S. Taipale. 2023. 'Individual and shared digital repertoires: Older adults managing digital services', *Information, Communication and Society* 26(3): 568–83. https://doi.org/10.1080/1369118X.2021.1954976.

Hartwick, J. and H. Barki. 1994. 'Explaining the role of user participation in information system use', *Management Science* 40(4): 440–65.

Heart, T. and E. Kalderon. 2013. 'Older adults: Are they ready to adopt health-related ICT?', *International Journal of Medical Informatics* 82(11): e209–e231. https://doi.org/10.1016/j.ijmedinf.2011.03.002.

Helbostad, J., B. Vereijken, C. Becker, C. Todd, K. Taraldsen, M. Pijnappels, K. Aminian and S. Mellone. 2017. 'Mobile health applications to promote active and healthy ageing', *Sensors* 17(3): 622. https://doi.org/10.3390/s17030622.

Helsper, E. 2016. 'Inequalities in digital literacy: Definitions, measurements, explanations and policy implications'. In *TIC domicílios 2015: Pesquisa sobre o uso das tecnologias de informação e comunicação nos domícilios brasileiros [ICT households 2015: Survey on the use of information and communication technologies in Brazilian households*], 175–85. Comitê Gestor da Internet no Brasil. https://eprints.lse.ac.uk/68329/.

Hill, R., L. R. Betts and S. E. Gardner. 2015. 'Older adults' experiences and perceptions of digital technology: (Dis)empowerment, wellbeing and inclusion', *Computers in Human Behavior 48*: 415–23. https://doi.org/10.1016/j.chb.2015.01.062.

Instituto Brasileiro de Georgrafia e Estatística (IBGE). 7 June 2023. *Pnad Contínua Educação 2022* ['*Continuous PNAD Education 2022*']. PowerPoint slides, 7 June 2023. Accessed 10 January 2025. https://agenciadenoticias.ibge.gov.br/media/com_mediaibge/arquivos/8100b5c6e47300b5b9596ced07156eda.pdf.

Junqueira, L., A. Freire, A. Grützmann, A. and E. Zitkus. 2023. 'Challenges and barriers faced by older adults to access legislative e-participation platforms in Brazil', *The Electronic Journal of Information Systems in Developing Countries* 89(5): e12274. https://doi.org/10.1002/isd2.12274.

Kickbusch, I., J. M. Pelikan, F. Apfel and A. D. Tsouros (eds). 2013. *Health Literacy: The solid facts*. World Health Organization. Accessed 10 January 2025. https://iris.who.int/handle/10665/326432.

van Laar, E., A. van Deursen, J. van Dijk and J. de Haan. 2017. 'The relation between 21st-century skills and digital skills: A systematic literature review', *Computers in Human Behavior* 72: 577–88. https://doi.org/10.1016/j.chb.2017.03.010.

Lindgren, I., C. Ø. Madsen, S. Hofmann and U. Melin. 2019. 'Close encounters of the digital kind: A research agenda for the digitalization of public services', *Government Information Quarterly* 36(3): 427–36. https://doi.org/10.1016/j.giq.2019.03.002.

Lloyds. 2021. *Essential Digital Skills Report 2021: Benchmarking the essential digital skills of the UK. Benchmarking the essential digital skills of the UK.* Third edition, 1 June 2021. Available upon request at DigitalSkillsInclusion@lloydsbanking.com.

Lombardi, P. and A. Vanolo. 2015. 'Smart city as a mobile technology: Critical perspectives on urban development policies'. In *Transforming City Governments for Successful Smart Cities. Public administration and information technology*, edited by M. Rodríguez-Bolívar, vol 8. Cham: Springer International Publishing. https://doi.org/10.1007/978-3-319-03167-5_8.

Miller, D., L. A. Rabho, P. Awondo, M. de Vries, M. Duque, P. Garvey, L. Haapio-Kirk, C. Hawkins, A. Otaegui, S. Walton and X. Wang. 2021. *The Global Smartphone: Beyond a youth technology*. London: UCL Press. https://doi.org/10.2307/j.ctv1b0fvh1.

Ministério da Economia. 2020. *Brasil lança sua Estratégia de Governo Digital para 2020 a 2022* [*Brazil launches its Digital Government Strategy for 2020 to 2022*]. Gov.Br. Accessed 29 April 2020. https://www.gov.br/economia/pt-br/assuntos/noticias/2020/abril/brasil-lanca-sua-estrategia-de-governo-digital-para-2020-a-2022.

Mobile Time and Opinion Box. 2021. Mensageria no Brasil - Agosto de 2021. Accessed 10 January 2025. https://www.mobiletime.com.br/pesquisas/mensageria-no-brasil-agosto-de-2021/.

Nedelcu, M. 2017. 'Transnational grandparenting in the digital age: Mediated co-presence and childcare in the case of Romanian migrants in Switzerland and Canada', *European Journal of Ageing* 14(4): 375–83. https://doi.org/10.1007/s10433-017-0436-1.

Niehaves, B. and R. Plattfaut. 2014. 'Internet adoption by the elderly: Employing IS technology acceptance theories for understanding the age-related digital divide', *European Journal of Information Systems* 23(6): 708–26. https://doi.org/10.1057/ejis.2013.19.

Nimrod, G. 2018. 'Technophobia among older internet users', *Educational Gerontology* 44(2–3): 148–62. https://doi.org/10.1080/03601277.2018.1428145.

Olson, K. E., M. A. O'Brien, W. A. Rogers and N. Charness. 2010. 'Diffusion of technology: Frequency of use for younger and older adults', *Ageing International* 36(1): 123–45. https://doi.org/10.1007/s12126-010-9077-9.

Olsson, T. and D. Viscovi. 2018. 'Warm experts for elderly users: Who are they and what do they do?', *Human Technology* 14(3): 324–42. https://doi.org/10.17011/ht/urn.201811224836.

Oudshoorn, N. 2011. *Telecare Technologies and the Transformation of Healthcare*. Houndmills, Basingstoke: Palgrave Macmillan.

Pan, S. and M. Jordan-Marsh. 2010. 'Internet use intention and adoption among Chinese older adults: From the expanded technology acceptance model perspective', *Computers in Human Behavior* 26(5): 1111–19. https://doi.org/10.1016/j.chb.2010.03.015.

Penido, A. 2018. *Estudo aponta que 75% dos idosos usam apenas o SUS* [*Study shows that 75% of the elderly use only SUS*]. 4 October 2018. Fiocruz. Accessed 10 January 2025. https://portal.fiocruz.br/noticia/estudo-aponta-que-75-dos-idosos-usam-apenas-o-sus#:~:text=O%20Elsi%2D%20Brasil%20apontou%20que,m%C3%A9dica%20nos%20%C3%BAltimos%2012%20meses.

Peral, B. P., J. A. Gaitán and M. Á Ramón-Jerónimo. 2014. 'Technology acceptance model y mayores: ¿la educación y la actividad laboral desarrollada son variables moderadoras?' ['Technology acceptance model and seniors: Are education and work activity moderating variables?'], *Revista Española de Investigación de Marketing ESIC* 18(1): 43–56. https://doi.org/10.1016/S1138-1442(14)60005-X.

Pesquisa Nacional por Amostra de Domicílios Contínua (PNAD). 2018. *Internet and television access and possession of mobile phone for personal use – National Continuous Household Sample Survey – Annual disclosure*. Brasilia: Brazilian Institute of Geography and Statistics (IBGE).

Pesquisa Nacional por Amostra de Domicílios Contínua (PNAD). 2019. '2019 Internet and television access and possession of mobile phone for personal use' (brochure). Brasilia: Brazilian Institute of Geography and Statistics (IBGE). Accessed 10 January 2025. https://www.ibge.gov.br/estatisticas/sociais/populacao/17270-pnadcontinua.html?edicao=30362.

Plaza, D. and L. Plaza. 2019. 'Facebook and WhatsApp as elements in transnational care chains for the Trinidadian diaspora', *Genealogy* 3(15): 55–74. https://doi.org/10.3390/genealogy3020015.

Poupatempo. 2022. *Dados estatíticos: Histórico de atendimentos desde 1997 a fevereiro de 2022* [*Statistical data: Service history from 1997 to February 2024*], Governo do Estado de São Paulo accessed 10 May 2022. https://www.poupatempo.sp.gov.br/wps/portal/poupatempoTaOn/o-poupatempo/dados-estatisticos.

Prado, A. M. 2016. 'O Jeitinho Brasileiro: Uma revisão bibliográfica' ['The Brazilian way: A literature review'], *Horizonte Científico* 10(1): 1–21.

Prado, M. M. and M. J. Trebilcock. 2018. *Institutional Bypasses: A strategy to promote reforms for development*. First edition. Cambridge, UK: Cambridge University Press. https://doi.org/10.1017/9781108563116.

Prodam. 2018. 'Aplicativo Agenda Fácil completa três meses de operação' ['Agenda Fácil App completes three months of operation'], online report, 8 February 2018. Accessed 10 May 2022. https://www.prefeitura.sp.gov.br/cidade/secretarias/inovacao/prodam/noticias/?p=249550.

Rede Nossa São Paulo. 2020. *Mapa da Desigualdade 2020* [*Inequality Map 2020*]. Instituto Cidades Sustentáveis. Accessed 10 January 2025. https://www.nossasaopaulo.org.br/wp-content/uploads/2020/10/Mapa-da-Desigualdade-2020-MAPAS-site-1.pdf.

Rivinen, S. 2021. *Developing Media Literacy Geragogy for Older People through Design-based Research*. PhD thesis, Faculty of Education, University of Lapland: Rovaniemi, Finland. https://urn.fi/URN:ISBN:978-952-337-288-7.

Rocha, E. and S. Padovani. 2016. 'Usability and accessibility in smartphones. Original title 'Usabilidade e Acessibilidade em Smartphones', *Ergodesign & Amp; HCI, 4*(especial): 58–66. https://dx.doi.org/10.22570/ergodesignhci.v4iEspecial.119.

Santana, S., B. Lausen, M. Bujnowska-Fedak, C. E. Chronaki, H.-U. Prokosch and R. Wynn. 2011. 'Informed citizen and empowered citizen in health: Results from an European survey', *BMC Family Practice* 12(1): 20. https://doi.org/10.1186/1471-2296-12-20.

Secretaria Municipal de Direitos Humanos e Cidadania & Coordenadoria de Políticas para Pessoa Idosa. 2019. *Indicadores Sociodemográficos da População Idosa na Cidade de São Paulo* [*Sociodemographic Indicators of the Elderly Population in the City of São Paulo*]. e-publication. Cidado de São Paulo. Accessed 13 March 2025. https://www.prefeitura.sp.gov.br/cidade/secretarias/upload/direitos_humanos/IDOSO/PUBLICACOES/Indicadores%20sociais_10_02_2020%20(3).pdf

Selwyn, N. 2004. 'The information aged: A qualitative study of older adults' use of information and communications technology', *Journal of Ageing Studies* 18(4): 369–84. https://doi.org/10.1016/j.jaging.2004.06.008.

SP News. 2020. 'Governo de São Paulo Lança Aplicativo "Poupatempo Digital"' ['Government of São Paulo launches "Poupatempo Digital" app'], 5 June 2020. Accessed 5 June 2020. https://www.saopaulo.sp.gov.br/noticias-coronavirus/governo-de-sp-lanca-aplicativo-poupatempo-digital/#:~:text=O%20Governador%20Jo%C3%A3o%20Doria%20anunciou%2C%20nesta%20quarta-feira%20%286%29%2C,sem%20necessidade%20de%20deslocamento%20a%20um%20posto%20f.

United Nations Population Division. 2022. *Probabilistic Population Projections based on the World Population Prospects 2022* (File PPP/POP60+). Dataset. Accessed 10 January 2025. https://population.un.org/wpp

van Deursen, A. and J. Van Dijk. 2011. 'Internet skills and the digital divide', *New Media & Society* 13(6): 893–911. https://doi.org/10.1177/1461444810386774.

van Dijk, J. and A. van Deursen. 2014. *Digital Skills: Unlocking the information society*. New York: Palgrave Macmillan US. https://doi.org/10.1057/9781137437037.

WBG GovTech Maturity Index. 2022. '2022 Update: Trends in public sector – Digital Transformation'. Washington, DC: World Bank. License: Creative Commons Attribution CC BY 4.0. Accessed 10 January 2025. https://documents1.worldbank.org/curated/en/099035001132365997/pdf/P1694820bcef0903e091160315d2050d03b.pdf.

World Health Organization (WHO). 2002. *Active Ageing: A policy framework*. (No. WHO/NMH/NPH/02.8). Accessed 10 January 2025. https://iris.who.int/handle/10665/67215.

Zitkus, E. and C. Libânio. 2021. 'Accessibility of the Brazilian public healthcare system during COVID-19 pandemic'. In *Reflexões Sobre Inclusão, Diversidade e Acessibilidade em Tempos de COVID-19*, edited by C. de S. Libânio and L. M. de Oliveira B. Silveira, 151–62. Porto Alegre: EditoraUFCSPA. https://www.ufcspa.edu.br/editora_log/download.php?cod=049&tipo=pdf.

7

Uncovering older adults' digital repertoires for designing conversational scenarios: graphic transcript as a design method

Sanna Kuoppamäki, Mikaela Hellstrand and Donald McMillan

Introduction

Conversational agents have become a part of everyday life, providing opportunities to act as collaborators and conversational partners in older adults' daily lives. These voice assistants are often not designed to meet the digital repertoires of older adults, however, which can result in a low adherence to these technologies from adults aged 65 and over. To ensure that conversational agents can be more adaptive to older adults' digital repertoires, recent developments in human-robot interaction aim to capture users' lived experiences, perceptions and interactions as part of their real-life environment – in which these technologies are co-produced in relation to the socio-material environment (Graf et al. 2024; Zawieska and Hannibal 2023; Veling 2017). This can help to push the field beyond mere acceptance towards a long-term user engagement, as well as providing a more appropriate and enjoyable design of conversational agents for various age groups (Veling and McGinn 2021).

This chapter investigates how graphic transcripts can inform the design process of conversational agents through uncovering older adults' digital repertoires as they unfold in their home environment. The study utilises video-ethnomethodology to inscribe the digital and non-digital repertoires as they are embedded in the sequences and temporalities of everyday life. We inscribe visual, spatial and temporal aspects of action through triangulation of text, visual annotations and video stills (Laurier 2014). We then investigate how digital repertoires can be made accessible through graphic transcripts, which can inform design scenarios of dialogue models through visualisation of different tasks, actions and linguistic styles. Such visualisations provide a resource to imagine a dialogue system with different conversational styles based on older adults' digital repertoires in a real-life environment. This in turn can ensure a more value-sensitive conversational model.

We consider the kitchen to be a space for exploring digital and non-digital repertoires, where technologies can facilitate older adults' autonomy and agency, improve the social connectedness between older adults and their family members and contribute to independent living and ageing in place (Kosch et al. 2018; Paay et al. 2015). Home cooking is a cognitive activity that consists of synchronising different recipe steps in a certain chronological order (Kuoppamäki et al. 2021). It is a more complex activity than conventional information seeking, for which conversational agents are traditionally used, and requires a high level of context-awareness from the agent to adapt to a user's situational and personal preferences. Digital repertoires can manifest themselves as routines, both individual and socially shared, reflecting the practices and habits of the household (Hänninen et al. 2023). Only by uncovering and adapting to these repertoires in their personal and social contexts can emerging technologies be considered as meaningful by the end-users themselves.

The field of human-robot interaction (HRI) engages researchers from several disciplines: computer science, robotics, engineering, human-computer interaction, psychology and social science. The common approach in HRI relies on its experimental settings that aim to produce generalisable, measurable and data-driven outcomes for the purpose of designing robots – rather than of developing a theoretical basis for interaction with robots. Experimental psychology conducted in experimental settings has been the gold standard (Dautenhahn 2018). However, what experimental work cannot address is how real people, in a real world setting, interact with real robots over a long period of time (Dobrosovestnova et al. 2024; Dautenhahn 2018; Zawiska and Hannibal 2023). The potential of an ethnomethodological approach in HRI is to re-imagine the relationship with participants and potential users of technology by using their insights, experience, domain expertise and creativity to apply to robot use and its possibilities. Through video-ethnomethodological analysis, we show how older adults' digital and non-digital repertoires can be made visible and accessible to inform engineers and designers. This can help to push the field of operation beyond mere acceptance. It also allows for the possibility of radical and transformative design towards a more inclusive, critical and reflexive stance with regard to wider societal and ethical concerns in the HRI field.

The chapter begins by introducing conversational agents for older adults and establishing the kitchen environment as a specific context for exploring digital repertoires. This is followed by an analysis and presentation of graphic transcripts as a methodological resource for video analysis. We develop design scenarios for conversational assistance and discuss ethical and practical concerns of anthropomorphising conversational agents in relation to older adults' digital repertoires. Research questions are the following:

1. How can graphic transcripts inform the design process of conversational agents?
2. What, if any, are the age-specific characteristics for designing conversational scenarios?

Conversational agents and older adults

Research on the usage of conversational agents (CA) has rapidly evolved, focusing mostly on the usage patterns, features and interactions in everyday life. These voice assistants have the potential to enhance the physical and cognitive interactions among older adults, both in terms of maintaining an active and healthy lifestyle and of augmenting various levels of care (Harrington and Egede 2023; Zubatiy et al. 2021; 2023; Wong et al. 2024; Yang et al. 2024). Research has addressed the usage of conversational agents and voice-based interfaces among older adults. These studies have focused on the usability and acceptability of conversational agents (Pradhan et al. 2019; Sayago et al. 2019), accessibility of voice interfaces (Blair and Abdullah 2019), usage patterns and satisfaction of use (Harrington and Egede 2023; Chung et al. 2019; Pradhan et al. 2020), new embodiments of conversational agents (El Kamali et al. 2020; Simpson et al. 2020), design requirements (Desai et al. 2023; Nikitina et al. 2018; Motta and Quaresma 2023), technical and linguistic requirements (Ding et al. 2022) and the application of conversational agents in caregiving (Zubatiy et al. 2021, 2023).

Older adults may perceive not only unique challenges but also benefits in the usage of conversational agents. Using voice as an interaction paradigm can help to overcome many of the difficulties that older adults typically experience with interactive technologies (Pradhan et al. 2020; Desai et al. 2023). Voice interaction should consider conversation content and linguistic style when responding to the experiences of this age group (Sayago et al. 2019). For instance, older adults with vision and mobility impairments can use voice interfaces to complete complex tasks (Blair and Abdullah 2019). However, older adults may not know how to speak with a conversational agent, at least initially (Sayago et al. 2019). Nor may older adults with hearing impairments have access to voice interaction (Desai et al. 2023). The linguistic content and default voices that conversational agents provide when interacting with older adults should therefore be designed appropriately from the conversational user experience, enabling it to support the more inclusive conversational agent (CA).

Older adults may also use conversational agents for specific purposes. Pradhan et al. (2020) investigated older adults' perceptions and usage patterns of intelligent voice assistants embodied in smart speakers, and noted the challenges associated with usage of voice assistants. Older adults used the voice-based smart speaker to seek online information about health, local businesses and food and drink. Most older adults perceived the voice interface as easier to use and learn compared with traditional computing devices, indicating that voice-based interfaces can improve the accessibility of digital technology. Supporting memory with reminders was less frequently used in comparison to searching for information. Older adults were concerned about the reliability of reminders, such as forgetting to set a reminder correctly and unstable internet connectivity. Other challenges in the usage of voice assistants included unpredictability and instability (for example, devices that timed out before the completion of voice commands), in addition to the lack of clarity and inconsistency regarding voice commands.

Conversational agents can also be leveraged to support the use of other online services. Voice-based augmentation (Sato et al. 2011) originally developed for blind users has the potential to reduce cognitive load, so can support older adults in using already familiar online services. Older adults who use a voice-based augmented interface for banking and online shopping mentioned increased confidence in using online services, reflecting the potential of voice-based interfaces for encouraging older adults to appropriate new forms of digital applications (Sato et al. 2011). Conversational user interfaces therefore have the potential to reduce many of the barriers to technology adoption that older adults frequently perceive with new devices (Pradhan et al. 2020; Chin et al. 2024).

Despite the benefits of using conversational agents to support healthy ageing and augmenting various levels of care, conversational user interfaces raise concerns about the human-like attributes displayed in many conversational agents. Voice-based interactions based on dialogue systems have a greater risk of anthropomorphising technology, defined as a tendency to attach human-like features to technical devices. When everyday technologies become 'anthropomorphised', users tend to attach personifying behaviours to these devices: they can be called 'friends' and 'someone to talk to' (Pradhan et al. 2019). Although anthropomorphism can increase the perceived likeability of voice assistants (Li and Sung 2021), human-like features of conversational agents can be leveraged to achieve emotional manipulation, such as to promote certain behaviours among users (Seymour and Van Kleek, 2021). Personification occurs especially when the voice is attached to a face or devices are called by the agent's name, or referred to as a personal pronoun, or when the dialogue system is based on polite greetings between humans and machines (Pradhan et al. 2019; Purington et al. 2017).

Voice assistants may also raise concerns over their 'always-on' nature that has no visual output (Lau et al. 2018; Pradhan et al. 2019). In the home, in general, privacy concerns include data collection (the type of data that is collected and the period for which it is retained) and the power dynamic between different household members (unequal access to smart home capabilities) (Lau et al. 2018). With voice assistants in particular, specific privacy concerns relate to speaking about private information in a socially shared location (whether the user needs to talk about health-related information in a socially shared location or not) and modality choice (whether the user needs to press a button to start voice interaction or not) (Moorthy and Vu 2013). Designing conversational agents for older adults should therefore acknowledge conversational style (Chin et al. 2024), linguistic content and different modality choices, as well as how they can facilitate both performance and connectivity in the kitchen.

The kitchen as a space for exploring digital repertoires

Conversational agents are becoming a part of the kitchen environment, as they have a greater ability to understand and predict user's information needs (Frummet et al. 2022, Kuoppamäki et al. 2023, Jaber et al. 2024). Technological support designed for kitchen activities has traditionally focused on instructional guidance and multimodal feedback on cooking, such as following recipes in the correct order. Hamada et al. (2005) used multimedia to offer instructional guidance on cooking based on the interpretation of cooking workflows and rescheduling recipe steps. Doman et al. (2012) offered instruction for each cooking step by means of personalised videos. Sato et al. (2013) designed a 'MimiCook' that displayed written instructions with video-projection to offer immediate feedback for the person in the kitchen.

Recent developments in the kitchen also include voice assistants and machine learning applications. For instance, Lim et al. (2019) used multimodal machine learning to develop effective recipe recommendations and minimise the overhead of tracking kitchen ingredients. Chang et al. (2019) investigated how cooking could be improved with instructional 'how-to' videos through a system that supports efficient content-based voice navigation through keyword-based queries. The system enables users to perform tasks with fewer commands and less frustration than in the conventional voice-enabled video interface.

Conversational agents also have a potential to support social and collaborative aspects of cooking, such as observing, checking, helping and showing (Paay et al. 2015). Our previous study (Kuoppamäki et al. 2021) showed that certain actions in the kitchen are challenging to complete in the time available, with respect to other tasks that also require attention. In temporal co-ordination of completing tasks, older adults typically rely on external help, such as that of family members, in managing the constraints of task completion (Kuoppamäki et al. 2021). Family members can assist in both temporal and physical tasks, providing either direct physical assistance or observing instances for external assistance.

Technologies that aim to improve the social and collaborative aspects of cooking have nevertheless been limited to cooking appliances that aim to share the experience of cooking together at a distance (Davis et al. 2014; Terrenghi et al. 2007). However, for older adults conversational agents could support the collaborative aspects of cooking. This co-ordination work could include the task division between older adults and their family members, the monitoring of ongoing tasks and the organisation of work (Kosch et al. 2018).

In human-computer interaction (HCI), digital repertoires translate to the recognition of older adults' needs, skills and abilities in relation to individualistic and collaborative tasks that, in the kitchen, can vary between organisation and reorganisation of tasks (Kuoppamäki et al. 2021). Family members and warm experts (Hänninen et al. 2020) can share and become a part of older adults' digital repertoires, by showing, checking, observing and helping in time-sensitive tasks (Kuoppamäki et al. 2021; Paay et al. 2015). Considering the kitchen as a space of daily activity for which conversational assistance could be developed indicates that technologies designed for older adults need to be adaptive to various levels of physical and cognitive ability. Such technologies should also offer a sense of personalisation and attachment through which they can improve the overall experience of cooking in the kitchen. Furthermore, these technologies have the potential to improve talk, communication and general social responsiveness between older adults and their family members, through which these technologies can mediate the social connectedness in everyday activities.

Data and methods

Designing technologies for kitchen activities requires cultural awareness and sensitivity towards the traditions and experiences of individuals (Dolejšová et al. 2018). For that reason we propose ethnography as a suitable method to investigate the everyday kitchen activities of a specific group of people, and the way that their behaviours are shaped by the systems and processes of which they are a part.

Ethnography and digital ethnography

Traditionally, ethnographic fieldwork has been utilised in long-term research projects, with an emphasis on theoretical, as well as conceptual, development and refinement (Hammersley and Atkinson 2019). However, when researchers from other disciplines make use of ethnography, it tends to be shaped in a way that makes it hard for ethnography to stay within its original boundaries and definition. In design research, the timeframe is often shorter and there is an emphasis on findings leading to applied interventions (Pink and Morgan 2013).

Several researchers have aimed to deconstruct ethnography to inform system designers. For instance, Button et al. (2015) developed a systematic method for building the social context into system design. The social context now occupies a significant place in design thinking, especially through collaboration with ethnomethodologists. Applying an ethnographic approach in design research means that a flexible approach towards theory and method should be maintained, since ethnography is largely about being adaptive and responsive to circumstances within the field.

Ethnomethodology and graphic transcript

One application area in digital ethnography is the graphic transcript originally introduced by Eric Laurier (2014). Graphic transcript originally derived from conversation analysis and ethnomethodology that aim to unpack the micro-level interactions between people and objects in a certain social situation. In comparison to traditional conversation analysis that investigates talk, gestures and communication with written transcripts, graphic transcript utilises visual images and comic book language to inscribe the visual, spatial and temporal aspects of action.

Comic-based research has recently gained popularity in qualitative research as a visual research method (Kuttner et al. 2021), but to the best of our knowledge, little or no research has explicitly shown how to apply this in design practice. The chapter's empirical focus is in investigating graphic transcripts as a methodological resource for video analysis and design. Our analysis shows the process of acquiring, applying and detecting the information (for example, physical interaction and temporalities) with graphic transcripts, which will then inform the development of design scenarios.

The benefit of graphic transcript in relation to traditional conversation analysis is the holistic approach to micro-level interactions, as it shows the continuity of interactions in a specific context. The limitations of this method lie in its ability to detect the user's lived experiences, since the method displays physical, temporal and social interaction as they are visible and observable from a naturalistic perspective – not from the phenomenological or experiential point of view.

Participants in the study

Empirical material presented in this study was collected among older adults (aged 65 and over) in their place of residence – a house or a flat where they either lived alone or with their spouse (Kuoppamäki et al. 2021). The data consist of video recordings of six sessions of participants cooking a meal, each approximately 60 minutes in length. Participants were recruited through their prior participation in design workshops at the university in Stockholm, Sweden. An invitation was sent via email to 26 participants, and seven replied to volunteer: they agreed to be recorded on video while preparing a meal in their kitchen.

Four of the participants were female and three were male. In preparation for the observations and recordings, participants were asked to make a meal of their own choice in their kitchen. We recorded the whole cooking session, from the preparation of ingredients to the plating of the finished meal. Video recordings started when the participants began to arrange objects in the kitchen (Kuoppamäki et al. 2021).

Analytical approach

As with all representations, the creation of graphic transcripts is a value-laden endeavour in which interpretations of action are presented to the reader through – and as a result of – the constraints of the medium. By 'value-laden endeavour', we mean that the selection of instances and vignettes are influenced by prior knowledge and understanding of the kitchen environment, the specific activities and tasks included, and the focus on older adults as a target group. Research on human-computer and human-robot interaction has typically relied on stereotypical images of older adults as frail or incompetent, in which technologies aim to 'fix' the problem of ageing (Lee and Riek 2018). Video-based ethnomethodology, and graphic transcripts in particular, allows the depiction of real bodies having real interactions that can decrease the impact of stereotypical images of older adults as a target group (Bath 2014).

The analytical procedure occurred in four steps (Kuoppamäki et al. 2021). First, we watched all the videos with the aim of forming a comprehensive overview of the content of the videos, such as the flow of actions, conflicts and interruptions in the cooking process. Second, in order to focus on age-specificity, we then extracted instances involving fluencies or disfluencies in action. Third, we created collections of short video clips (between approximately 1 and 3 minutes in length) which we collectively analysed in data analysis sessions. To protect the participants' privacy they were given pseudonyms, their faces were blurred in the images and all forms of personal details were removed. From the collection of video clips, we finally selected vignettes, actions and interactions to highlight and to fall between the panels of the comic. For use in further analysis, multiple graphic transcripts can be created to be used in analysis sessions; this can be done iteratively and collaboratively. For user-centred design, the representations can be chosen to expose either interactional possibilities or challenges to be overcome by design.

With graphic transcripts we analyse both multimodality and sequence and simultaneity of action. Multimodality refers to presenting data and analysis simultaneously by creating tension between the images and words to highlight ambiguity, uncertainty and multiple perspectives.

It opens up opportunities to take analysis in divergent directions, to build up multiple layers of meaning and to explore themes that do not necessarily fall into a traditional model of causality. During analysis, the creation of page spreads affords a chance to lay out diverse pieces of data from across a study, such as drawings from field notes, participant-created images, photographs and quantitative data. This can be used to explore similarities, contrasts, patterns and an overall intuitive sense of what is emerging from the research.

Older adults' repertoires in the kitchen

Video recording offers a possibility for interpreting and analysing kitchen activities. One can make use of the medium of video by pausing and rewinding to explore physical and temporal repertoires among older adults. This offers greater clarity than through participant observation, as the mundane and the multimodal are difficult to see in a real-life situation, in which several things are occurring simultaneously. The kind of mundane practices that comprise digital repertoires – things that are not even thought about, but are embodied as multi-tasking skills – form part of our everyday activities and the bodily know-how or habitus (Hindmarsh and Tutt 2012; Martens 2012). Video-based ethnomethodology can be employed to expose and make digital repertoires available for designing particular scenarios.

Physical and spatial interaction

The benefits and affordances of graphic transcripts over textual descriptions and video screenshots derive from their ability to communicate both physical and spatial interaction in a more holistic way than via traditional conversation analysis (Laurier 2014; 2019). Physical interaction can be demonstrated in comics by the use of movement lines, arrows and highlights attached to the image. By addressing movements in the space, we can expose the physicality in relation to the space, and in connection to different objects such as tools and artefacts in the kitchen. An example of this is the vignette of bending down to reach the bottom of the fridge (Fig. 7.1). This example shows how kitchen activities are influenced by various interruptions and disfluencies. Some of them are associated with age-related capabilities while others are caused by contextual factors and the kitchen environment as a physical space.

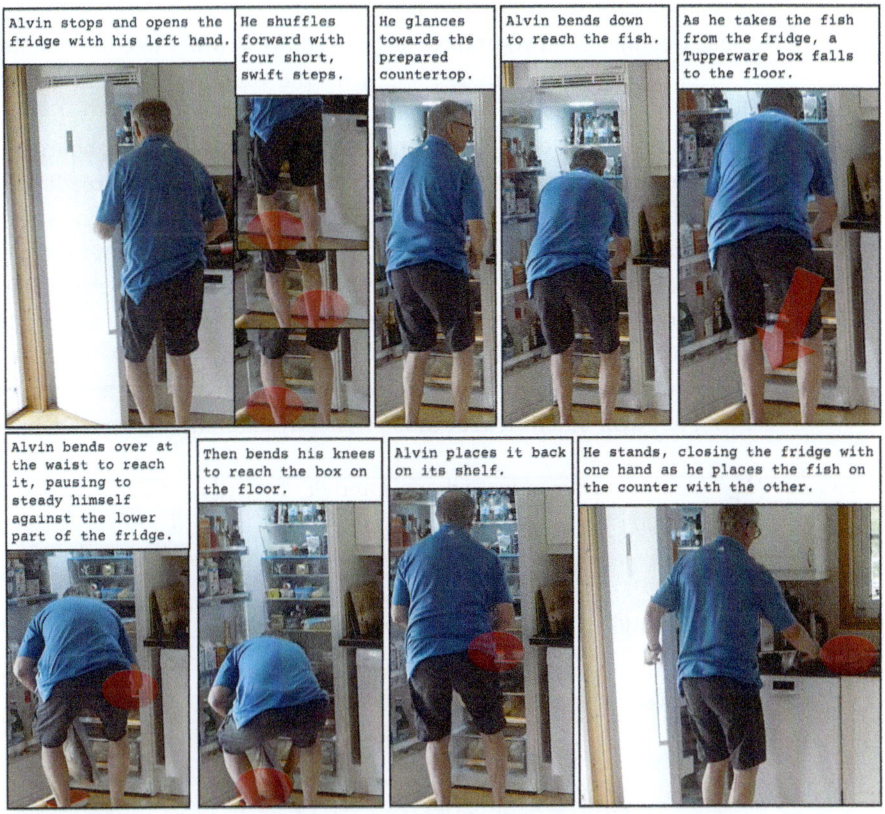

Figure 7.1 A graphic transcript of physical interactions and movement in the space. Image by Sanna Kuoppamäki. Edited by Donald McMillan.

Social and collaborative tasks

Graphic transcripts can also address the sociality and physicality of sharing space and tasks. In the kitchen there are often bottlenecks where resources are required for two ongoing tasks at the same time. Accomplishing both tasks together involves negotiating space for the body and access to the tools and ingredients involved. This negotiation often requires the ordering of tasks, taking into account their relative time sensitivity and the social aspects of who should be doing what and when.

Clara and her husband Benny engage in such a negotiation around access to the sink without a single word spoken (Fig. 7.2). In this vignette, Benny has begun to scrub the potatoes just as Clara has finished preparing the patty mixture for the meatballs. As he finishes washing the first potato, he realises that the water in the pot is the water that was used to soak the unwashed vegetables rather than fresh water to boil them in. He turns up the water pressure and, with the washed potato still in his other hand, proceeds to empty, rinse and fill the pot with fresh water. At the same time as he does this, Clara has brought the chopping board on which she plans to form the meatballs to be rinsed. She holds it ready just on the edge of the sink as a visible token of her intention to use the sink.

Upon finishing his task of filling the pan, Benny does not turn the flow of water down to continue washing the potatoes or leave the flow high for the task of cleaning the board. Instead he turns the water off completely – misreading the intent in Clara's proximity and the object she was holding. Clara then reaches across the sink, turning on the tap and leaving her arm outstretched, thus physically claiming the space while

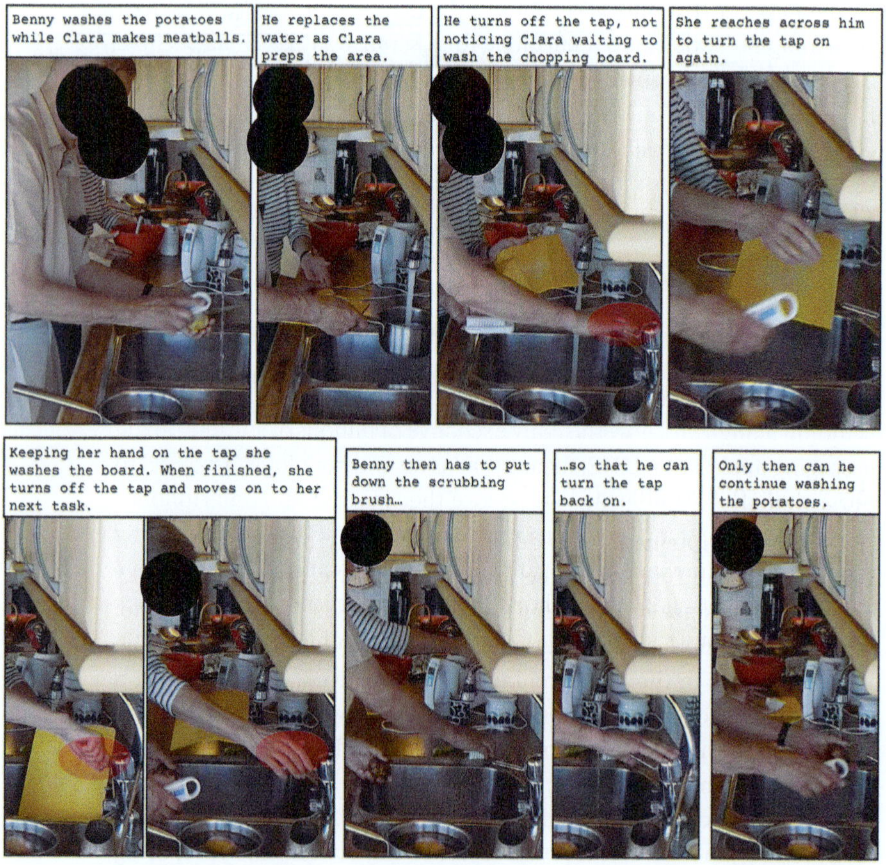

Figure 7.2 A graphic transcript of social and collaborative tasks. Image by Sanna Kuoppamäki. Edited by Donald McMillan.

she is using it. Benny then waits, potato and scrubber at the ready, making clear his own intent to continue the washing process. Upon finishing her cleaning of the board, Clara turns off the tap and turns back to her task. With both hands full and his intent to use the running water misread, Benny then has to put down the brush to turn on the water and continue towards his goal of scrubbed and boiled potatoes.

Temporalities and interconnected actions

Laurier (2019) looks in detail at the representation of complex, interconnected streams of talk and action in the graphic transcript format. The temporalities expressed in this format are, by their nature, the sequentiality of action rather than the timing of action. As can be seen in the visual transcript (Fig. 7.3), this format offers a connection between the next and previous actions as captured in the strip – but it does not provide the exact timings that a traditional conversation analysis transcript would. While it would be possible to add timings to the captions – for example, borrowing from Jefferson notation to add to caption 3 in Figure 7.3 to indicate how long an action took ('The chef stops pouring (after 1.2 seconds) and rests the mixing bowl (3.8 seconds) until the butter is hot enough') – this could never give a complete picture of the timing of actions within a sequence. The time between panels in this format is only indicated in terms of deviation from the implied sequentiality of each cell. Overlapping or connected cells indicate quickly following actions (such as the shuffle forward shown in Figure 7.1), parallel actions or long gaps with phrases of time such as 'Meanwhile…' or 'Sometime later…'.

Figure 7.3 A graphic transcript of temporal interaction and interconnected actions. Image by Sanna Kuoppamäki. Edited by Donald McMillan.

This focus on the sequentiality and physicality of action lends itself to the analysis of interactions that go beyond conversational (Fig. 7.2). Where conversation is presented, this focuses the researcher on how talk and action are interwoven, rather than on the internal relationships between parts and participants of speech.

Designing scenarios for conversational assistance

Graphic transcripts can be used to inform the design process, and the creation of graphic transcripts to demonstrate design decisions. In our work, the graphic transcripts presented above were made available as a resource for technology design (Crabtree et al. 2012). Designing scenarios were thus influenced by the creation of graphic transcripts: the information of the physical, social and temporal aspects of action, and the fluencies and disfluencies revealed in certain cooking tasks. In addition, the participatory observation, in the form of a collection of video recordings, provided the researcher with knowledge and understanding of older adults' home cooking process.

Building on the combination of this information, we developed 20 different scenarios for conversational assistance in the kitchen. These scenarios were focused on three actions: tool suggestion, orientation of action and recipe reminding. Two conversational agents, Furhat (Moubayed et al. 2012) and Tama (McMillan et al. 2019), were employed with two different linguistic styles: direct command and conversational suggestion. Furhat and Tama are both conversational AI platforms that use speech and gaze as a main interaction modality.

Tool suggestion

One challenge older adults frequently perceived in the kitchen was the selection and maintenance of appropriate tools. People sometimes visibly struggled to chop onions, for example, which can in part be explained by a knife's relative bluntness or in one instance by a participant's limited strength, reflected in her own statement that arthritis had made such actions increasingly painful and difficult in recent years (Kuoppamäki et al. 2021). This situation could therefore be supported by a conversational assistant aimed at tool suggestion (Fig. 7.4). For this scenario, we employed the following scripts: 'Hey chef! Maybe use the mandolin?'[1] (conversational) and 'Tool suggestions! A mandolin would be a good tool for this job'[2] (direct).

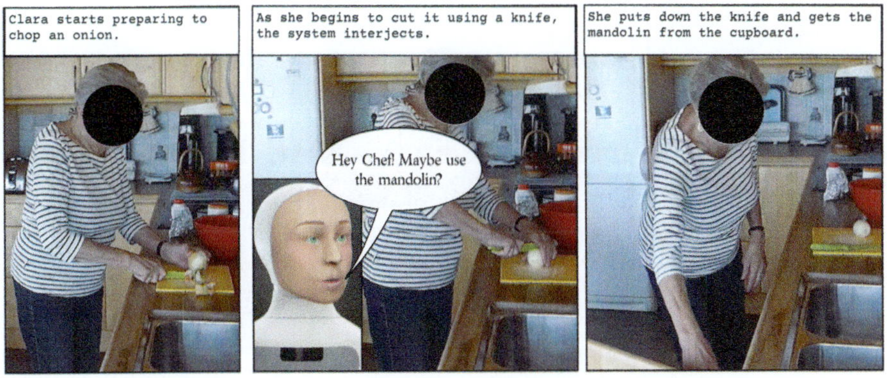

Figure 7.4 A design scenario of tool suggestion. Image by Sanna Kuoppamäki. Edited by Donald McMillan.

Recipe reminding

Sometimes older adults did not follow the recipe in the correct order. This was demonstrated as reorganisation of kitchen tasks, for example inserting additional tasks or temporal co-ordination in completing tasks (Kuoppamäki et al. 2021). Transitional moments of moving from one sub-task to another can elicit a re-evaluation of cooking tasks, especially during moments of temporal constraints. For recipe ordering and following, we envisioned a recipe reminder with the following scripts: 'Hey Chef! Did you remember Taco Spice?'[3] (conversational) and 'You skipped the Taco Spice. Not needed?'[4] (direct) (Fig. 7.5).

Figure 7.5 A design scenario of recipe reminding. Image by Sanna Kuoppamäki. Edited by Donald McMillan.

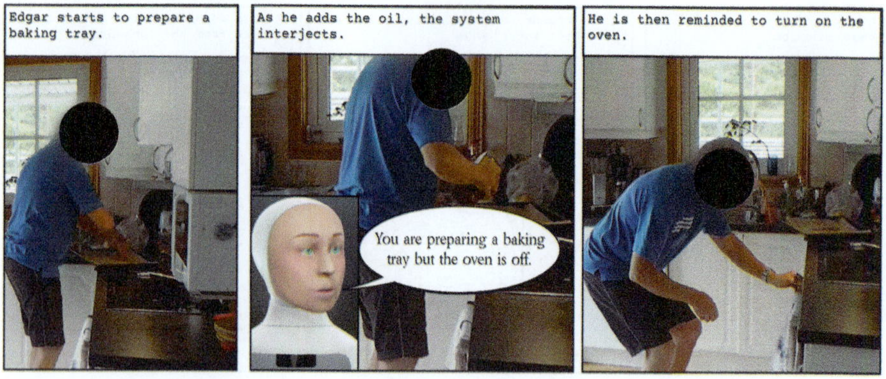

Figure 7.6 A design scenario of orientation of action. Image by Sanna Kuoppamäki. Edited by Donald McMillan.

Orientation of action

Our previous study also showed that cooking involves conducting several tasks simultaneously, requiring a level of organisational fluency to switch between, monitor and co-ordinate tasks and to organise tools and objects in the kitchen space (Kuoppamäki et al. 2021). Physical and cognitive capabilities may change with age, especially in relation to hearing and vision, working memory and attention span. While ageing may be accompanied by decreasing mobility, knowledge of familiar recipes and routines may help to compensate somewhat for any physical limitations (Kuoppamäki et al. 2021). The process could nevertheless be supported by conversational assistance aiming to direct attention towards certain recipe steps that are prone to be confused or forgotten. This could be implemented in conversational scripts, such as 'You are preparing a baking tray, but the oven is off'[5] (conversational) and 'Hey chef! The oven is not turned on!'[6] (direct) (Fig. 7.6).

Designing conversational scenarios with digital repertoires

This chapter has investigated how graphic transcripts can uncover older adults' digital repertoires to design conversational scenarios for independent living. Building on the findings from our previous video study on older adults' home cooking (Kuoppamäki et al. 2021), the chapter has highlighted the process of using video-ethnomethodology and graphic transcripts as a specific approach and method that can inform the design process of conversational agents, by focusing on the physical and spatial interactions and temporalities in the kitchen. We have then presented design scenarios focused on three actions: tool suggestion, orientation of action and recipe reminding. Two conversational agents and two different linguistic styles were employed: direct command and conversational suggestion. Here we discuss the potential of these scenarios in relation to linguistic style, anthropomorphisation of conversational agents and designing with older adults' digital repertoires.

Linguistic style

Conversational user interfaces can reduce many barriers to new technology adoption that older adults frequently perceive (Pradhan et al. 2020). However, the conversational style can influence the perceived likeability of the agent in later life (Chin et al. 2024). Studies have recognised that linguistic style should be customisable, suggesting that aesthetic aspects of voice interaction are related to age (Chin et al. 2024). Other aspects of linguistic style are well documented, such as hearing loss, vocabulary and idiom sets among older adults (Blair and Abdullah 2020). Other studies note that older adults are either faster or slower to interact with voice and speech interaction (Stigall et al. 2019).

Age-specific features of voice-based interaction systems have mostly been recognised from the perspective of technical requirements. Gollasch and Weber (2021) identified age-specific strategies in dialogue systems and speech recognition accuracy. To respond to the needs of older adults, voice assistants should be able to recognise even unusual formulations correctly; complex dialogues comprising multiple pieces of information should be presented as simple or guided dialogues; voice assistants should ask only one question per dialogue with a limited set of possible answers; voice assistants should be able to retain information about the conversation context, and so on. However, many confounding factors such as experience with voice user interface (VUI) technology, and technology in general, as well as interest in the outcome of using the device, could also play a more important role with older adults (Pradhan et al. 2019; 2020).

Although aesthetic aspects of voice and speech interaction may not be connected to chronological age as such, as a specific media repertoire they may entail features connected to certain time periods. Furthermore, older adults' preferences for certain media landscapes have shown to be generational, i.e. connected to one's formative youth period (Bolin 2015; Ivan et al. 2020). People often form a passionate and nostalgic relationship with certain musical genres and stars, as well as reproduction technologies (not only for vinyl records, but also for music cassette tapes and comics) prevalent during the formative youth period.

Such passion can take many forms. In audience studies, a large body of work concerns fans and other media users who are passionately engaged, seeking pleasure in and receiving empowerment from the certain media landscape (Bolin 2015). To the best of our knowledge, no previous studies have raised the questions of generational identity or experience in relation to the usability or likeability of conversational agents. It could be argued, however, that these questions become more relevant when conversational agents are to be developed as aesthetic and pleasant assistants with respect to audio and speech interaction, possessing the ability to detect and respond not only to the physical, but also to the emotional states of users.

Anthropomorphisation of conversational agents

Conversational agents embodying interaction modalities such as voice, gaze and other human-like attributes are subject to anthropomorphisation, defined as a tendency to attach human-like features to artificial agents. Anthropomorphisation can increase the acceptance of conversational agents to a certain extent, as users tend to evaluate human-like agents more positively in comparison to those with non-human-like features. Attaching human-like attributes to conversational agents has shown to increase trust, improve the interaction and perceived sociability of CAs (Li and Sung 2021). However, this appraisal also depends on the task and context (Sin and Muntenau 2019). Our study has presented scenarios for conversational assistance mediated by a conversational platform in three activities: tool suggestion, recipe reminding and orientation of action. We perceive that relying on conversational rather than direct physical assistance in a private environment like home could be a more appropriate approach for designing assistance that respects older adults' capabilities without seeing ageing as 'a problem to be fixed with technology' (Lee and Riek 2018). Supporting physical and cognitive functioning with gaze interaction can respond to older adults' autonomy and agency, since these modalities can respond to older adults' technological capability in a less disruptive way and thus minimise the perception of assistive nature of these technologies (Bright and Coventry 2013).

Our study has not yet validated the scenarios nor investigated older adults' responses and reactions to the anthropomorphisation of conversational agents. However, it has contributed to recognition of relevant questions that could be covered in future research. These questions include for instance: How do older adults, and their family members or friends, also referred as warm experts, react and respond to different interaction modalities that display human-like attributes of CAs? To what extent are the connection between anthropomorphisation of CAs and the perceived likeability of CAs associated with age? What are the age-specific features of CA usage that go beyond technical aspects towards socially shared practices of voice interfaces? To answer these questions, our future study will aim to investigate both individual and socially shared repertoires for different conversational systems in a real-life environment based on the scenarios developed in this study.

Designing with older adults' digital repertoires

Digital repertoires, a concept derived from media repertoires and the individual lifestyle of the person in the social context (Schwarzenegger 2020), has been proposed as a solution to conceptualise older adults' digital technology use in relation to personal interests, skill sets and the availability of digital devices and applications (Hänninen et al. 2020). Digital repertoires describe both the individual and shared use of digital technologies, arguing that digital technology use should go beyond an individual perspective to encompass the everyday use of digital technology, extending to social practices and shared meanings ascribed to digital technologies (Hänninen et al. 2023). Sharing can be a one-directional practice from a 'warm expert' to an older adult and indicate the older adult's dependency on external assistance in technology use. Yet it can also be a genuinely reciprocal process. For instance, social support shapes the connection between digital and social inclusion through immediate social networks: 'warm experts' in particular serve as a key source of emotional, informational and instrumental support that enhances the access to and use of digital technologies in later life (Kuoppamäki et al. 2022). Voice-based interaction systems designed to support older adults' independent living should thus respond to the shared practices between older adults and their warm experts, by facilitating talk, communication and emotional responsiveness between older adults and the warm expert.

Designing new interaction systems that better respond to the needs and skill sets of older adults can also indicate changes between the roles of warm experts and older adults. Previous studies have addressed the various levels of support that warm experts provide for older adults in their digital inclusion and technology use (Hänninen et al. 2020). Warm experts help older adults identify the skills or situations with which they most need assistance by providing an opportunity for social comparisons. They provide a source of social attachment that increases older adults' own interest and motivation for digital engagement; and they provide both technical and non-technical assistance that can compensate for the challenges older adults experience in the course of ageing (Kuoppamäki et al. 2022). Conversational user interfaces, displaying a higher level of adaptability to user's needs, have the potential to reduce technology adoption barriers among older adults (Pradhan et al. 2020). Therefore the significance of warm experts in providing social support in digital inclusion may decrease, as older adults obtain greater abilities to manage digital technologies on their own.

Conclusions

In this chapter we have presented an original design method that makes visible the individual, socially shared and contextual aspects of digital repertoires that often remain unrecognised or undervalued in design practice. Digital repertoires can be translated to manifold ways of adhering with digital technologies in their wider socio-material environment, and to the individual's skills and motivations as they unfold in sequences and temporalities of actions. Traditionally these aspects of repertoires have been difficult to visualise, capture or concertise within social science methodologies, which has resulted in perceiving 'everyday life' as being of secondary importance in fields such as human-robot interaction (Zawieska and Hannibal 2023). It has thus been difficult to take digital repertoires into account in robot design.

By incorporating video-ethnomethodology to visualise the digital and non-digital repertoires highlighted in Chapter 1, however, we can produce more actionable visualisations. These provide a common and accessible language of annotation of the complex temporal relationships between action and interaction in a real-world technology use. They also provide a triangulation through their combination of text, visual annotations and video stills to render complex physical and social interactions between people, technology and the wider environment less ambiguous.

For engineers and designers, our approach provides a novel way of learning from users, and of recognising the complexity and unpredictability of users' actions in certain contexts or environments through text, visual annotations and video stills. Social scientists can use this approach to illustrate digital repertoires in a tangible manner, thus making embedded and everyday practices more visible and concrete. Our collaborative studies have shown that this can result in new interaction paradigms to imagine alternative ways of using social robots in later life (Kuoppamäki et al. 2023; Jaber et al. 2024). Creating user representations of digital repertoires as they unfold in their authentic environment can challenge the existing, often stereotypical, images of ageing. This in turn has the potential to bring about a more value-sensitive approach in design fields.

In the present volume both this chapter and Chapter 5 (by Laura Haapio-Kirk) use graphic visual methods to explore digital repertoires. In so doing they demonstrate how integrating visual approaches can bridge disciplines such as design and technology studies, sociology and anthropology. A visual approach can foster greater interdisciplinary collaboration, encouraging scholars and practitioners from various fields to engage with the study of digital repertoires in new and innovative ways.

Acknowledgements

The research is funded by Digital Futures, KTH (project: Advanced Adaptive Intelligent Systems).

Notes

1 Hej kocken! Kanske funkar mandolin?
2 Verktygsförslag! Mandolin skulle vara en bra redskap för det här jobbet!
3 Hej kocken! Har du kommit ihåg kryddorna?
4 Krydden hoppar du över. Onödigt!
5 Du använder en bakplåt, men ugnen är inte på uppvärmning.
6 Hej kocken! Ugnen är inte igång.

References

Bath, C. 2014. 'Searching for methodology: Feminist technology design in computer science'. In *Gender in Science and Technology: Interdisciplinary Approaches*, edited by W. Ernst and I. Horwath, 57–78. Bielefeld: Transcript Verlag.

Blair, J. and S. Abdullah. 2019. 'Understanding the needs and challenges of using conversational agents for deaf older adults'. [Poster presentation]. *Conference Companion Publication of the 2019 on Computer Supported Cooperative Work and Social Computing*, 161–5. https://doi.org/10.1145/3311957.3359487.

Blair, J. and S. Abdullah. 2020. 'It didn't sound good with my cochlear implants: Understanding the challenges of using smart assistants for deaf and hard of hearing users', *Proceedings of the ACM on Interactive, Mobile, Wearable and Ubiquitous Technologies* 4(4): 1–27. https://doi.org/10.1145/3432194.

Bolin, G. 2015. 'Passion and nostalgia in generational media experiences', *European Journal of Cultural Studies* 19(3): 250–64. https://doi.org/10.1177/1367549415609327.

Bright, A. K. and L. Coventry. 2013. 'Assistive technology for older adults: Psychological and socio-emotional design requirements', *Proceedings of the 6th International Conference on PErvasive Technologies Related to Assistive Environments* (PETRA '13), 1–4. https://doi.org/10.1145/2504335.2504344.

Button G., A. Crabtree, M. Rouncefield and P. Tolmie. 2015. *Deconstructing Ethnography. Towards a Social Methodology for Ubiquitous Computing and Interactive Systems Design*. Cham, Switzerland: Springer International Publishing. https://doi.org/10.1007/978-3-319-21954-7.

Chang, M., A. Truong, O. Wang, M. Agrawala and J. Kim. 2019. 'How to design voice based navigation for how-to videos'. *Proceedings of the 2019 CHI Conference on Human Factors in Computing Systems* 701, 1–11. https://doi.org/10.1145/3290605.3300931.

Chin, J., S. Desai, S. Lin, H. Cheng and S. Mejia. 2024. 'Like my Aunt Dorothy: Effects of conversational styles on perceptions, acceptance and metaphorical descriptions of voice assistants during later adulthood', *Proceedings of the ACM on Human-Computer Interaction 8(CSCW1)*, 1–21. https://doi.org/10.1145/3637365.

Chung, K., Y. H. Oh and D. Y. Ju. 2019. 'Elderly users' interaction with conversational agent'. [Poster presentation]. *Proceedings of the 7th International Conference on Human-Agent Interaction*, 277–9. https://doi.org/10.1145/3349537.3352791.

Crabtree, A., M. Rouncefield and P. Tolmie. 2012. *Doing Design Ethnography*. London: Springer.

Dautenhahn, K. 2018. 'Some brief thoughts on the past and future of human-robot interaction'. *ACM Transactions on Human-Robot Interaction* 7(1): 1–3. https://doi.org/10.1145/3209769.

Davis, H., B. Nansen, F. Vetere, T. Robertson, M. Brereton, J. Durick and K. Vaisutis. 2014. 'Homemade cookbooks: A recipe for sharing', *Proceedings of the 2014 Conference on Designing Interactive Systems* 73–82. https://doi.org/10.1145/2598510.2598590.

Desai, S., X. Hu, M. Lundy and J. Chin. 2023. 'Using experience-based participatory approach to design interactive voice user interfaces for delivering physical activity programs with older adults', *International Conference on Human-Agent Interaction*, 180–90. https://doi.org/10.1145/3623809.3623827.

Ding, Z., J. Kang, T. O. T. Ho, K. H. Wong, H. H. Fung, H. Meng and X. Ma. 2022. 'TalkTive: A conversational agent using backchannels to engage older adults in neurocognitive disorders screening', *CHI Conference on Human Factors in Computing Systems*, 1–19. https://doi.org/10.1145/3491102.3502005.

Dobrosovestnova, A., H. R. Lee, S. Ljungblad, M. Gamboa, T. Gosnall and M. Mansouri. 2024. 'Ethnography in HRI: Embodied, embedded, messy and everyday', *Companion of the 2024 ACM/IEEE International Conference on Human-Robot Interaction*, 1314–16. https://doi.org/10.1145/3610978.3638547.

Dolejšová, M., R. A. Khot, H. Davis, H. S. Ferdous and A. Quitmeyer. 2018. 'Designing recipes for digital food futures', *Extended Abstracts of the 2018 CHI Conference on Human Factors in Computing Systems* 1–8. https://doi.org/10.1145/3170427.3170622.

Doman, K., C. Y. Kuai, T. Takahashi, I. Ide and H. Murase. 2012. 'Smart VideoCooKing: A multimedia cooking recipe browsing application on portable devices', *Proceedings of the 20th ACM International Conference on Multimedia* (MM '12), 1267–8. https://doi.org/10.1145/2393347.2396435.

El Kamali, M., L. Angelini, M. Caon, O. A. Khaled, E. Mugellini, N. Dulack, P. Chamberlin, C. Craig and G. Andreoni. 2020. 'NESTORE: Mobile chatbot and tangible vocal assistant to support older adults' wellbeing', *Proceedings of the 2nd Conference on Conversational User Interfaces* 34, 1–3. https://doi.org/10.1145/3405755.3406167.

Frummet, A., D. Elsweiler and B. Ludwig. 2022. '"What can I cook with these ingredients?" Understanding cooking-related information needs in conversational search', *ACM Transactions on Information Systems* 40(4): 1–32. https://doi.org/10.1145/3498330.

Gollasch, D. and G. Weber. 2021. 'Age-related differences in preferences for using voice assistants', *Mensch Und Computer 2021*, 156–67. https://doi.org/10.1145/3473856.3473889.

Graf, P., C. S. Zarp-Falden, L. Naik, K. B. Lefeuvre, E. Marchetti, E. Hornecker, M. B. Sørensen, L. V. Hemmingsen, E. V. Just Christensen, L. Bodenhagen, N. Krüger and A. Bischof. 2024. 'Distributed agency in HRI – an exploratory study of a narrative robot design', *Frontiers in Robotics and AI* 11, 1253466. https://doi.org/10.3389/frobt.2024.1253466.

Hamada, R., J. Okabe, I. Ide, S. Satoh, S. Sakai and H. Tanaka. 2005. 'Cooking navi: Assistant for daily cooking in kitchen', *Proceedings of the 13th Annual ACM International Conference on Multimedia* (MULTIMEDIA '05), 371–4. https://doi.org/10.1145/1101149.1101228.

Hammersley, M. and P. Atkinson. 2019. *Ethnography: Principles in practice* (fourth edition). London: Routledge. https://doi.org/10.4324/9781315146027.
Hänninen, R., S. Taipale and R. Luostari. 2020. 'Exploring heterogeneous ICT use among older adults: The warm experts' perspective', *New Media & Society* 23(6): 1584–1601. https://doi.org/10.1177/1461444820917353.
Hänninen, R., L. Pajula, V. Korpela and S. Taipale. 2023. 'Individual and shared digital repertoires – older adults managing digital services', *Information, Communication & Society* 26(3): 568–83. https://doi.org/10.1080/1369118X.2021.1954976.
Harrington, C. N. and L. Egede. 2023. 'Trust, comfort and relatability: Understanding Black older adults' perceptions of chatbot design for health information seeking', *Proceedings of the 2023 CHI Conference on Human Factors in Computing Systems*, 1–18. https://doi.org/10.1145/3544548.3580719.
Hindmarsh, J. and D. Tutt. 2012. 'Video in analytic practice'. In *Advances in Visual Methodology*, edited by S. Pink, 57–73. London; Los Angeles, CA: SAGE Publications Ltd.
Ivan, L., E. Loos and I. Bird. 2020. 'The impact of "technology generations" on older adults' media use: Review of previous empirical research and a seven-country comparison', *Gerontechnology* 19(4): 1–19. https://doi.org/10.4017/gt.2020.19.04.387.
Jaber, R., S. Zhong, S. Kuoppamäki, A. Hosseini, I. Gessinger, D. P. Brumby, B. R. Cowan and D. McMillan. 2024. 'Cooking with agents: Designing context-aware voice interaction', *Proceedings of the CHI Conference on Human Factors in Computing Systems,* 1–13. https://doi.org/10.1145/3613904.3642183.
Kosch, T., P. W. Woźniak, E. Brady and A. Schmidt. 2018. 'Smart kitchens for people with cognitive impairments: A qualitative study of design requirements', *Proceedings of the 2018 CHI Conference on Human Factors in Computing Systems*, 1–12. https://doi.org/10.1145/3173574.3173845.
Kuoppamäki, S., S. Tuncer, S. Eriksson, S. and D. McMillan. 2021. 'Designing kitchen technologies for ageing in place: A video study of older adults' cooking at home', *Proceedings of the ACM on Interactive, Mobile, Wearable and Ubiquitous Technologies* 5(2): 1–19. https://doi.org/10.1145/3463516.
Kuoppamäki, S., R. Hänninen and S. Taipale. 2022. 'Enhancing older adults' digital inclusion through social support: A qualitative interview study'. In *Vulnerable People and Digital Inclusion: Theoretical and applied perspectives,* edited by P. Tsatsou, 211–230. Cham: Palgrave Macmillan. https://doi.org/10.1007/978-3-030-94122-2_11.
Kuoppamäki, S., R. Jaberibraheem, M. Hellstrand and D. McMillan. 2023. 'Designing multimodal conversational agents for the kitchen with older adults: A participatory design study', *International Journal of Social Robotics* 15(9–10): 1507–23. https://doi.org/10.1007/s12369-023-01055-4.
Kuttner, P. J., M. B. Weaver-Hightower and N. Sousanis. 2021. 'Comics-based research: The affordances of comics for research across disciplines', *Qualitative Research* 21(2): 195–214. https://doi.org/10.1177/1468794120918845.
Lau, J., B. Zimmerman and F. Schaub. 2018. '"Alexa, are you listening?": Privacy perceptions, concerns and privacy-seeking behaviors with smart speakers', *Proceedings of the ACM on Human-Computer Interaction* 2(CSCW): 1–31. https://doi.org/10.1145/3274371.
Laurier, E. 2014. 'The graphic transcript: Poaching comic book grammar for inscribing the visual, spatial and temporal aspects of action', *Geography Compass* 8(4): 235–48. https://doi.org/10.1111/gec3.12123.
Laurier, E. 2019. 'The panel show: Further experiments with graphic transcripts and vignettes', *Social Interaction. Video-Based studies of human sociality* 2(1). https://doi.org/10.7146/si.v2i1.113968.
Lee, H. R. and L. D. Riek. 2018. 'Reframing assistive robots to promote successful ageing', *ACM Transactions on Human-Robot Interaction* 7(1): 1–23. https://doi.org/10.1145/3203303.
Li, X. and Y. Sung. 2021. '"Anthropomorphism brings us closer": The mediating role of psychological distance in User–AI assistant interactions', *Computers in Human Behavior* 118: 106680. https://doi.org/10.1016/j.chb.2021.106680.
Lim, H., X. Huang, S. Miller, J. Edelmann, T. Euken and S. Voida. 2019. 'Smart cook: Making cooking easier with multimodal learning'. [Poster presentation]. *Adjunct Proceedings of the 2019 ACM International Joint Conference on Pervasive and Ubiquitous Computing and Proceedings of the 2019 ACM International Symposium on Wearable Computers*, 129–32. https://doi.org/10.1145/3341162.3343836.

Martens, L. 2012. 'The politics and practices of looking: CCTV video and domestic kitchen practices'. In *Advances in Visual Methodology*, edited by S. Pink, 39–56. London; Los Angeles, CA: SAGE Publications Ltd.

McMillan, D., B. Brown, I. Kawaguchi, R. Jaber, J. Solsona Belenguer and H. Kuzuoka. 2019. 'Designing with gaze: Tama, a gaze activated smart-speaker', *Proceedings of the ACM on Human-Computer Interaction* 3(CSCW): 1–26. https://doi.org/10.1145/3359278.

Moorthy, A. E. and K.-P. L. Vu. 2013. *Voice Activated Personal Assistant: Privacy concerns in the public space*. Los Angeles, CA: California State University. https://www.proquest.com/docview/1513579796/abstract/DFF130DF16554E2FPQ/1.

Motta, I. and M. Quaresma. 2023. 'Increasing transparency to design inclusive conversational agents (CAs): Perspectives and open issues', *Proceedings of the 5th International Conference on Conversational User Interfaces*, 1–4. https://doi.org/10.1145/3571884.3604304.

Moubayed, Al S., J. Beskow, G. Skantze and B. Granström. 2012. 'Furhat: A back-projected human-like robot head for multiparty human-machine interaction'. In *Proceedings of the 2011 International Conference on Cognitive Behavioural Systems*, edited by A. Esposito and A. M. Esposito, 114–30. Berlin; Heidelberg: Springer. https://doi.org/10.1007/978-3-642-34584-5_9.

Nikitina, S., S. Callaioli and M. Baez. 2018. 'Smart conversational agents for reminiscence', *Proceedings of the 1st International Workshop on Software Engineering for Cognitive Services*, 52–7. https://doi.org/10.1145/3195555.3195567.

Paay, J., J. Kjeldskov and M. B. Skov. 2015. 'Connecting in the kitchen: An empirical study of physical interactions while cooking together at home', *Proceedings of the 18th ACM Conference on Computer Supported Cooperative Work & Social Computing*, 276–87. https://doi.org/10.1145/2675133.2675194.

Pink, S. and J. Morgan. 2013. 'Short-term ethnography: Intense routes to knowing', *Symbolic Interaction* 36(3): 351–61. https://doi.org/10.1002/symb.66.

Porcheron, M., J. E. Fischer, S. Reeves and S. Sharples. 2018. 'Voice interfaces in everyday life', *Proceedings of the 2018 CHI Conference on Human Factors in Computing Systems*, 1–12. https://doi.org/10.1145/3173574.3174214.

Pradhan, A., L. Findlater and A. Lazar. 2019. '"Phantom friend" or "just a box with information": Personification and ontological categorization of smart speaker-based voice assistants by older adults', *Proceedings of the ACM on Human-Computer Interaction* 3(CSCW): 1–21. https://doi.org/10.1145/3359316.

Pradhan, A., A. Lazar and L. Findlater. 2020. 'Use of intelligent voice assistants by older adults with low technology use', *ACM Transactions on Computer-Human Interaction* 27(4): 1–27. https://doi.org/10.1145/3373759.

Purington, A., J. G. Taft, S. Sannon, N. N. Bazarova and S. H. Taylor. 2017. '"Alexa is my new BFF": Social roles, user satisfaction and personification of the Amazon Echo', *Proceedings of the 2017 CHI Conference Extended Abstracts on Human Factors in Computing Systems*, 2853–9. https://doi.org/10.1145/3027063.3053246.

Sato, D., M. Kobayashi, H. Takagi, C. Asakawa and J. Tanaka. 2011. 'How voice augmentation supports elderly web users', *The Proceedings of the 13th International ACM SIGACCESS Conference on Computers and Accessibility* (ASSETS '11), 155–62. https://doi.org/10.1145/2049536.2049565.

Sato, A., K. Watanabe and J. Rekimoto. 2013. 'MimiCook: A cooking assistant system with situated guidance', *Proceedings of the 8th International Conference on Tangible, Embedded and Embodied Interaction* (TEI '14), 121–4. https://doi.org/10.1145/2540930.2540952.

Sayago, S., B. B. Neves and B. R. Cowan. 2019. 'Voice assistants and older people: Some open issues', *Proceedings of the 1st International Conference on Conversational User Interfaces* (CUI '19), 1–3. https://doi.org/10.1145/3342775.3342803.

Schwarzenegger, C. 2020. 'Personal epistemologies of the media: Selective criticality, pragmatic trust and competence–confidence in navigating media repertoires in the digital age', *New Media & Society* 22(2): 361–77. https://doi.org/10.1177/1461444819856919.

Seymour, W. and M. Van Kleek. 2021. 'Exploring interactions between trust, anthropomorphism, and relationship development in voice assistants', *Proceedings of the ACM on Human-Computer Interaction* 5(CSCW2): 1–16. https://doi.org/10.1145/3479515.

Simpson, J., F. Gaiser, M. Macík and T. Breßgott. 2020. 'Daisy: A friendly conversational agent for older adults', *Proceedings of the 2nd Conference on Conversational User Interfaces* (CUI '20), 1–3. https://doi.org/10.1145/3405755.3406166.

Sin, J. and C. Munteanu. 2019. 'A preliminary investigation of the role of anthropomorphism in designing telehealth bots for older adults'. In *Extended Abstracts of the 2019 CHI Conference on Human Factors in Computing Systems* (CHI EA'19), 1–6. https://doi.org/10.1145/3290607.3312941.

Stigall, B., J. Waycott, S. Baker and K. Caine. 2019. 'Older adults' perception and use of voice user interfaces: A preliminary review of the computing literature'. In *Proceedings of the 31st Australian Conference on Human-Computer-Interaction*, 423–7. https://doi.org/10.1145/3369457.3369506.

Terrenghi, L., O. Hilliges and A. Butz. 2007. 'Kitchen stories: Sharing recipes with the living cookbook', *Personal and Ubiquitous Computing* 11(5): 409–14. https://doi.org/10.1007/s00779-006-0079-2.

Veling, L. 2017. 'Becoming real: An anthropological approach to evaluating robots in the real world'. In *Social Robotics*, edited by A. Kheddar, E. Yoshida, S. S. Ge, K. Suzuki, J.-J. Cabibihan, F. Eyssel and H. He, vol. 10652, 638–46. Cham, Switzerland: Springer International Publishing. https://doi.org/10.1007/978-3-319-70022-9_63.

Veling, L. and C. McGinn. 2021. 'Qualitative research in HRI: A review and taxonomy', *International Journal of Social Robotics* 13(7): 1689–1709. https://doi.org/10.1007/s12369-020-00723-z.

Wong, N., S. Jeong, M. Reddy, C. A. Stamatis, E. G. Lattie and M. Jacobs. 2024. 'Voice assistants for mental health services: Designing dialogues with homebound older adults', *Designing Interactive Systems Conference*, 844–58. https://doi.org/10.1145/3643834.3661536.

Yang, Z., X. Xu, B. Yao, E. Rogers, S. Zhang, S. Intille, N. Shara, G. G. Gao and D. Wang. 2024. 'Talk2Care: An LLM-based voice assistant for communication between healthcare providers and older adults', *Proceedings of the ACM on Interactive, Mobile, Wearable and Ubiquitous Technologies* 8(2): 1–35. https://doi.org/10.1145/3659625.

Zawieska, K. and G. Hannibal. 2023. 'Towards a conceptualisation and critique of everyday life in HRI', *Frontiers in Robotics and AI*, 10. https://doi.org/10.3389/frobt.2023.1212034.

Zubatiy, T., K. L. Vickers, N. Mathur and E. D. Mynatt. 2021. 'Empowering dyads of older adults with mild cognitive impairment and their care partners using conversational agents', *Proceedings of the 2021 CHI Conference on Human Factors*, 1–5. https://doi.org/10.1145/3411764.3445124.

Zubatiy, T., N. Mathur, L. Heck, K. L. Vickers, A. Rozga and E. D. Mynatt. 2023. '"I don't know how to help with that": Learning from limitations of modern conversational agent systems in caregiving networks', *Proceedings of the ACM on Human-Computer Interaction* 7(CSCW2): 1–28. https://doi.org/10.1145/3610170.

8

Epilogue

Daniel Miller

From hype to habit

There is a major problem with the study of digital technologies. However, this volume is an excellent example of a solution to that problem. The problem is that the digital attracts a different kind of academic response from almost any other topic. If we were studying history, chemistry, religion or class divisions, for example, we would expect a relatively gradual build-up of scholarship that extends from its foundations and develops new insights. Yet when it comes to the study of digital technologies, there seems to be an overwhelming desire to focus on whatever seems to be the 'next big thing'. As has been the case for some years now, I frequently hear from people who want to engage in some aspect of digital anthropology; right now the overwhelming proportion of these want to study some aspect of AI. Last year it might have been the study of data and algorithms, the year before that research on platformisation or perhaps TikTok. What makes things worse is that many of these researchers will not even concentrate on the consequences of these digital advances, as we can observe them. Instead they want to use their studies to predict the future – which is, of course, what most commercial firms are particularly interested in.

The result of this approach is bound to have a deleterious impact upon scholarship. Instead of consolidating and building, it seems like every year we start again. Together with my colleagues, I spent a considerable amount of time studying Facebook when it was the dominant social media platform. As anthropologists, we tried to use this example to make many general insights into the nature of social media platforms. Of course, it is right that people have subsequently moved on to study contemporary platforms. Attention should shift to TikTok. But what is dispiriting is that the new studies of the new platforms seem to make little use of what was learned from the previous studies of older platforms, even though many of the insights would remain relevant. Consider, for example, how social media relates to our sense of place, or extends forms of friendship and other relationships. This tendency to ignore the precedents and focus only on the present and prediction is reinforced by the media and popular interest, which have the exact same tendencies.

If that is the problem, then the solution would be scholarship directed to building up more fundamental approaches and observations that can be sustained even when the digital forms themselves continue to evolve and change. That is what makes this volume about digital routines particularly welcome. It represents a shift in attention from hype to habit. This volume is trying to excavate some fundamental properties that would apply almost irrespective of which digital technology we are concerned with. The chapters address inequalities that constantly reinvent themselves, the digital divide represented by differences between the young and older users (which only grows with additional life expectancy) and the ways in which our studies can influence policy and design. Above all, digital repertoires can apply to almost any engagement between a given population and whatever digital technologies they are in a position to use. This is what studies of the digital need to be.

The repertoire – digital and other

The key term in introducing this book is that of repertoire. The work of sociologist Elizabeth Shove (Shove 2003; Shove et al. 2012) reminds us that this term pertains to almost everything in life, not just the digital. Shove is particularly concerned with issues of energy use and climate change. She shows how the normative foundation for routines depends upon underlying conventions of comfort. What temperature do people normally set their thermostats to because that is assumed to be comfortable? How many layers of clothes do they wear inside or outside the house? Do they assume they should keep their windows open or closed? People in London may have entirely different attitudes to all three of these norms than people who live in Helsinki. These tend not to be questioned because comfort is typical of the unquestioned norm – but changing these norms may be central to any serious saving of energy among these populations.

Within the first hour of the day, we discover repertoires in the time someone gets up, what they eat for breakfast and how they prepare it, whether they find out about the news or not, how they wash, clean their teeth and get dressed. Everything is generally routine. Lofgren (2023), who has also written extensively about the role of routines and 'doing nothing' in everyday life, suggests that:

> Morning routines are about staggering from sleep to waking and getting body and mind ready for the outside world. Mornings are vulnerable times and routines become an important survival strategy; they consist of a number of small tasks that have to be coordinated, sequenced or multitasked (Lofgren 2023, 29).

This is fully reflected in Chapter 7 by Kuoppamäki, Hellstrand and McMillan. In this the authors use the kitchen as the space for exploring digital repertoires, just as Gullestad (1984) did for social routines around the kitchen.

There is an important caveat here. One of the contributions of this volume is the decision to use the word 'repertoire', while the authors I have just introduced are more concerned with routine. Choosing 'repertoire' – as emphasised in the introduction to this book, as well as in chapters by Ivan and others – expresses a preference for a more inclusive term that recognises not just what people routinely do, but also provides a context to encompass all that they may be capable of doing routinely. The colloquial meaning of 'repertoire' includes what a person has 'in their back pocket' – i.e. that which they are capable of, if need be. This is exactly the right quality to focus on in a book which includes a concern with how we might enable people to discover things within their personal repertoire and then to deploy them – precisely in order to go beyond what had become their normative routines.

In a way, 'digital repertoire' becomes the other side of the coin to a concept I developed with Madianou called 'polymedia' (Madianou and Miller 2012). Polymedia was intended to focus on the diversity of platforms available to users that creates an ecology, where each platform has meaning in relation to the other. As a result, the choice of platform came to be judged on moral and other grounds. However, the idea of a digital repertoire starts from the other end. Here the concern is with the combination of techniques used by the individual, or that they are capable of deploying, and the way in which each of these take their meaning from their complementarity to others. In short, this volume shows how individuals also have their own ecologies.

The use of the word repertoire, that is, recognising what people are capable of, seems essential to the excellent points made by Ivan and supported by Duque and Zitkus. This is important, not least because the very semantics of terms such as 'skills' and 'digital divide' can become problematic when they exacerbate the very divisions they are intending to address. Our starting point should be what older people themselves wish to achieve with digital technologies, not a top down 'schoolteacher' set of requirements they are expected to fulfil. We therefore need to start from people's wider lives and to consider their aspirations for the things that really matter to them, such as connecting with family and friends or being able to access services. This implies that we need to think of digital skills more as means than ends in themselves. Mostly it should not be about whether people have a particular digital skill, but the role of that skill in helping them to accomplish a task that matters to them.

That would be the preferred focus. Yet again there is a caveat here, as implied by Chapter 2 (Loredana Ivan) and by other chapters in this volume. If older people are denigrated for their inability to deal with the digital in itself, irrespective of what they plan to use the digital *for*, then that problem of prejudice has to be addressed in its own right. As several of the chapters in this volume demonstrate, a key finding from close observation is the propensity towards anxiety. If you do not fully understand the medium of communication, you may not be sure that the message conveyed is the one you intended. Was that the right emoji? Was it the most polite or the clearest way of expressing something online?

As Duque and Zitkus show, when it comes to health services people may prefer creatively to adapt the apps they are comfortable with, such as WhatsApp, rather than deploy specialist bespoke health apps. This is in part because WhatsApp is also the platform that facilitates their social connections, which in turn may be the key to better health treatment. The team of people I have worked with in a project on ageing and smartphones describe the ingenuity and creativity of patients and other users by the term 'smart-from-below' (Hawkins, Awondo and Miller 2024). We have then to acknowledge the issue of respect connected to whether older people feel they can use digital technologies or are intimidated by them.

The term repertoire is also valuable in the way it deals with another separate issue in considering the use of the digital by older people. Ideally our focus should really be on what it is that these people actually want to achieve – in which case it is not a question of whether they use some prescribed app or method for doing this. A better approach is to start from Hänninen and Taipale's general investigation of the use of the digital within daily lives, family and social networks. As Cabalquinto shows, this focus on people's actual goals makes the concern with context even more central to our research, because the context itself can radically change. What people needed and wanted was suddenly transformed by Covid-19, for example – when the very meaning and experience of co-presence become central to the basic sense of sociality.

Another of the significant contributions made by this volume is its emphasis upon the multimodal, and the need to give equal weight these days to non-textual elements that the internet has made a more significant part of human communication. One of the consequences of this shift, explored in Chapter 4 by Cabalquinto, is that this in turn leads to greater attention to the affective dimension. This is because what we find in exploring the multimodal is that different senses tend to evoke different kinds of emotional engagement. The digital means we are actually *seeing* the baby's first smile, rather than just reading about it, and we expect that to elicit a stronger emotional engagement.

These issues are exacerbated when one is talking about a migrant community (Ahlins 2023; Walton 2021). As Wang and Haapio-Kirk (2021) note, graphic forms, such as stickers in Japan or short videos in China, have become instrumental to the way in which people now express care. Yet equally care can be demonstrated through sending money for people of low income, for whom a health issue may be also a financial crisis – something very clear in the case of mobile money used in low-income Uganda (Hawkins 2023). Haapio-Kirk adds to this literature an attention to the role of these media in self-care. The observation that non-textual genres can broaden our modes of expression has carried through to methodology.

This is a major leitmotif of these volume. We see it in how Hänninen and Taipale use photo-elicitation, or the way in which Haapio-Kirk uses comic-making as elicitation and more general multimodal collaboration, while Kuoppamäki, Hellstrand and McMillan use what they call 'graphic transcripts' and 'video-ethnomethodology'. As a result, we have gained key and original insights into what the visual and auditory can convey beyond that of the textual.

It was also helpful that the book ends with Chapter 7 by Kuoppamäki, Hellstrand and McMillan because they bring out two points likely to become ever more important. The development of the multimodal provides a breadth which increasingly corresponds to what we see as an approximation to human beings – that is, engagements that include all our senses. This issue of anthropomorphism has accelerated greatly in the last couple of years thanks to AI and the increasing possibility of bringing these together in effective replication of persons, whether designing a 'partner', as in platforms such as *replika* and *otome* games, or 'bringing back' a deceased loved one.

The second point is that Kuoppamäki, Hellstrand and McMillan bring out more directly the responsibility of using their research to think more clearly not just about the interventions that might assist people in independent living, but the potential problematic and intrusive nature of digital interventions. This makes an appropriate finale to the book, because the major contribution of the idea of digital repertoires is the need always to retain the central role of empathy when dealing with individuals. We may consider them in their social context, their institutional context and the affordances of the technologies. Yet ultimately the idea of repertoire is to understand the goals of these individuals, and the range of resources they are capable of bringing to bear on trying to achieve them.

'The Anthropology of Smartphones and Smart Ageing' project and broader contexts

One of the reasons this volume is successful is that it is very focused – not only empirically, in terms of the problems it addresses, but also on its potential contribution to welfare. All of this is welcome, suggesting in turn a final role for an epilogue: to consider the still broader context that necessarily lies beyond the scope of this book. I hope you will forgive me if I spend a little time considering the contribution of our own work in showing how such a wider context might help in the further considerations that this book leads us towards. I will do this first by rethinking the meaning of our basic terms, and then by showing how this leads to an appreciation of the advantages of a comparative perspective.

From 2017 to 2022 a team of anthropologists, including myself and two of the authors in this volume, Marília Duque and Laura Haapio-Kirk, worked on the ASSA project, that is the Anthropology of Smartphones and Smart Ageing (Miller et al. 2021). As a large-scale comparative project, we also tried to consider the foundational assumptions behind our topic of study. We began with ageing. Over many volumes we were able to explore some of the main changes in the nature and experience of ageing. These varied hugely from contexts such as Japan, where Haapio-Kirk met plenty of very active 70, 80 and even 90-year-olds, to Hawkins' work on low-income populations in Kampala, where people may feel elderly from the age of 40. Going beyond the now accepted categories of 'third age' and 'fourth age' (Higgs and Gilleard 2017), several of the studies, including the one I carried out in Ireland with Garvey (Garvey and Miller 2021), noticed a fundamental shift. The populations we studied tended no longer to see themselves in relation to categories of age, but rather in continuity with youth, ending only with evident frailty, rather than any defined chronology. We would hope that these findings are also valuable in contextualising the discussions of the current volume. After all, a consideration of digital repertoires of ageing implicates our understanding of how ageing itself varies and is changing.

The second part of this relationship between the digital and ageing is our understanding of what we mean by the term 'digital'. Currently I am working (with Xinyuan Wang) on an edited volume concerned with digital anthropology in China. One of our major points is a significant difference in how people understand the implications of this term. In Europe the digital immediately became perceived as a dualism between the virtual and real. The digital has also been viewed in essence as a form of technology. We tend to take these for granted. So it is something of a shock to realise that in China the digital was not seen to contribute to this fundamental dualism in anything like the same extent.

This is because, over millennia, technology itself has been seen as something very different. Technology was above all an aspect of the more general project whereby the emperor was mandated by heaven to ensure the maintenance of social harmony. We argue that this remains an important foundation for how digital technologies are perceived in China today. This means that just as our assumptions about ageing are parochial, so are our assumptions about the digital.

Both of these observations derived from one particular facet of our project. ASSA was essentially a comparative study composed of multiple field sites around the world. Many of the subsequent insights derive from this explicit attention to the comparative. Having considered the meaning of ageing and the meaning of the digital, the core of the present volume concerns what happens when we focus on the relationship between ageing and the digital. An underlying assumption in most of the chapters in this volume has been an association between ageing and conservatism, with several chapters such as that by Ivan (Chapter 2) noting the problematic implications of this association. However, in her book about China, Wang (2023) shows that older people may strive to be in the vanguard of digital usage because they identify with the government project of using the digital to become the world's leading nation. This is important because it returns us to one of the benefits of that term repertoire: the fact that this term implies something more than the observation of how the digital relates to ageing. A repertoire is what people might be capable of.

In this case, the work on China helps us to understand that many of the limitations of older people in using the digital are not intrinsic. Rather they derive from historical conditions and expectations, including those prejudices referred to above.

This points to a final virtue of the present volume, which is the range of places the chapters address. There is clear continuity between the chapter in this volume by Haapio-Kirk (Chapter 5) and the arguments within her monograph (2024). The latter traces the enduring legacy of traditional gender categories, such that even today some Japanese men tend to remain loyal to the feature (*garakai*) phone they use (or formerly used) at work, while women more readily take up smartphones to develop their social connections in later life. This breadth of geographical and cultural foci underscores the present volume's capacity to reveal how deeply rooted cultural practices, such as those in Japan, shape and are shaped by technological adoption, illustrating the nuanced interplay between tradition, gender norms and the evolving landscape of digital communication.

The presence of a multitude of geographical locations in this volume invites questions that most discourse on ageing and digitalisation often fails to address. For example, are welfare goals general or localised? I have used a brief consideration of the ASSA project as movement from very broad questions about culture and terms to how we should address specific contexts and ultimately individuals. This discussion is intended to complement the present volume that in some ways starts with individuals and their routines, then gradually broadens out to address wider contextual issues.

This then brings us full circle to the start of my discussion. And that is appropriate. For the main way in which basic cultural assumptions become embedded and unquestioned is when they take the form of normative routines of the kinds investigated by Shove (2003) and Bourdieu (1977). Such routines in turn underlie the approach taken in this volume to digital repertoires.

Acknowledgements

The ASSA project discussed in this chapter was funded by the European Research Council (ERC) under the European Union's Horizon 2020 research and innovation programme (grant agreement No. 740472).

References

Ahlins, T. 2023. *Calling Family*. New Brunswick, NJ: Rutgers University Press.

Bourdieu, P. 1977. *Outline of a Theory of Practice*. Cambridge: Cambridge University Press.

Garvey, P. and D. Miller. 2021. *Ageing with Smartphones in Ireland: When life becomes craft*. London: UCL Press.

Gullestad, M. 1984. *Kitchen-Table Society*. Oxford: Oxford University Press.

Haapio-Kirk, L. 2024. *Ageing with Smartphones in Japan: Care in a visual digital age*. London: UCL Press.

Hawkins, C. 2023. *Ageing with Smartphones in Uganda: Togetherness in the dotcom age*. London: UCL Press.

Hawkins, C. P. Awondo and D. Miller (eds). 2024. *An Anthropological Approach to mHealth*. London: UCL Press.

Higgs, P. and C. Gilleard. 2017. *Re-thinking Old Age. Theorising the fourth age*. London: Bloomsbury.

Lofgren, O. 2023. 'Routines, rituals and reflexes: The powerful undercurrents in everyday life'. In *Against the Background of Social Reality*, edited by C. Lombardo and L. Sabetta, 26–41. Abingdon, Oxon: Routledge.

Madianou, M. and D. Miller. 2012. *Migration and New Media: Transnational families and polymedia*. Abingdon, Oxon: Routledge.

Miller, D., L. Abed Rabho, P. Awondo, M. de Vries, M. Duque, P. Garvey, L. Haapio-Kirk, C. Hawkins, A. Otaegui, S. Walton and X. Wang. 2021. *The Global Smartphone: Beyond a youth technology*. London: UCL Press. https://doi.org/10.14324/111.9781787359611.

Shove, E. 2003. *Comfort, Cleanliness and Convenience: The social organization of normality*. Oxford: Berg.

Shove, E., M. Pantzar and M. Watson. 2012. *The Dynamics of Social Practice: Everyday life and how it changes*. Los Angeles, CA: Sage.

Walton, S. 2021. *Ageing with Smartphones in Urban Italy: Care and community in Milan and beyond*. London: UCL Press.

Wang, X. 2023. *Ageing with Smartphones in Urban China: From the cultural to the digital revolution in Shanghai*. London: UCL Press.

Wang, X. and L. Haapio-Kirk. 2021. 'Emotion work via digital visual communication: A comparative study between China and Japan', *Global Media and China* 6(3): 325–44. https://doi.org/10.1177/20594364211008044.

Index

References to figures are in *italics*.